MTTC 020 Earth Space Science

Fern Harrison

This page is intentionally left blank.

This page is intentionally left blank.

Table of Content

Chapter 1 – Questions 1

Chapter 2 – Answers and Explanations 75

This page is intentionally left blank.

Chapter 1 – Questions

QUESTION 1

Which of the following best defines a scientific investigation?

    A.   A process of verifying preconceived ideas through experimentation.
    B.   A systematic approach to solving a problem using observations, data collection, and analysis.
    C.   A method of proving hypotheses to be true beyond a doubt.
    D.   A random collection of experiments conducted without a specific goal.

QUESTION 2

What is the primary purpose of organizing and analyzing scientific data?

    A.   To make the data look presentable and visually appealing.
    B.   To manipulate the data to support a preconceived conclusion.
    C.   To identify patterns, trends, and relationships within the data.
    D.   To minimize the amount of data collected during an investigation.

QUESTION 3

Why is it essential to present scientific data in a clear and unbiased manner?

    A.   To impress colleagues and supervisors with presentation skills.
    B.   To hide any flaws or limitations in the data collection process.
    C.   To ensure the data can be easily understood and evaluated by others.
    D.   To manipulate the data to fit personal beliefs and opinions.

QUESTION 4

When conducting a scientific investigation, what role do hypotheses play?

    A.   Hypotheses are the final conclusions drawn from the data analysis.
    B.   Hypotheses are the data collected during the investigation.
    C.   Hypotheses are the initial predictions or educated guesses that guide the research.
    D.   Hypotheses are the methods used to collect scientific data.

QUESTION 5

What is the evidential basis of scientific claims?

    A.   Emotional appeal and rhetorical arguments presented by researchers.
    B.   The reputation and credentials of the scientists making the claims.
    C.   The logical and empirical support from systematic observations and data.
    D.   The number of experiments conducted, regardless of the results.

QUESTION 6

Which of the following is NOT a criterion for collecting scientific data?

    A.   Data should be collected in a controlled environment.
    B.   Data should be collected using proper instruments and tools.
    C.   Data should be collected from a single source to ensure consistency.
    D.   Data should be collected with minimal bias and errors.

QUESTION 7

During a scientific investigation, what is the role of a control group?

    A.   To receive the experimental treatment to test the hypothesis.
    B.   To serve as a reference point for comparison with the experimental group.
    C.   To ensure the research is conducted without any ethical concerns.
    D.   To collect additional data that is not part of the main study.

QUESTION 8

Which of the following is NOT a step in the scientific method?

    A.   Formulating a hypothesis.
    B.   Drawing conclusions based on intuition.
    C.   Conducting experiments or data collection.
    D.   Analyzing the data and drawing conclusions.

QUESTION 9

When interpreting scientific data, why is it crucial to consider potential sources of bias?

    A.   To manipulate the data to support personal beliefs.
    B.   To ensure the data is consistent with previous research.
    C.   To account for any factors that may influence the results.
    D.   To simplify the data analysis process.

QUESTION 10

What is the purpose of peer review in the scientific process?

    A.   To promote competition among researchers.
    B.   To validate scientific claims without questioning.
    C.   To identify any errors or weaknesses in the research.
    D.   To provide funding for scientific investigations.

QUESTION 11

What is the primary mechanism responsible for plate tectonics and the movement of Earth's lithospheric plates?

    A.   Convection currents in the mantle
    B.   Solar radiation and gravitational forces
    C.   Earth's magnetic field
    D.   Friction between the crust and the atmosphere

QUESTION 12

Which of the following is a characteristic feature of an oxbow lake?

    A.   A crescent-shaped body of water formed by glacial erosion.
    B.   A circular depression resulting from a collapsed cave or cavern.
    C.   A U-shaped valley carved by a glacier.
    D.   A U-shaped lake formed when a meandering river is cut off from its main channel.

QUESTION 13

Which type of rock is formed by the cooling and solidification of magma or lava?

    A.   Sedimentary rock
    B.   Metamorphic rock
    C.   Igneous rock
    D.   Fossilized rock

QUESTION 14

What is the main cause of coastal erosion?

    A.   Human activities such as construction and mining
    B.   Earthquakes and volcanic eruptions
    C.   Wind and rain erosion
    D.   The action of waves and currents on the shoreline

QUESTION 15

What is the process by which a glacier breaks off large blocks of ice, creating icebergs in the ocean?

    A.   Calving
    B.   Ablation
    C.   Creep
    D.   Sublimation

QUESTION 16

Which of the following is a correct safety procedure for Earth and space science investigations?

    A.   Conduct experiments without safety goggles
    B.   Perform experiments alone without supervision
    C.   Dispose of chemicals in the regular trash
    D.   Wear appropriate personal protective equipment (PPE) such as safety goggles, lab coat, and gloves when conducting experiments.

QUESTION 17

Which of the following materials should be handled with care due to its potential hazards during Earth and space science investigations?

    A.   Paper
    B.   Plastic
    C.   Mercury
    D.   Aluminum foil

QUESTION 18

Which of the following is an appropriate mathematical procedure for reporting data in Earth and space science?

    A.   Rounding off all measurements to the nearest whole number
    B.   Presenting data without any labels or units
    C.   Using scientific notation for large or small numbers
    D.   Using random symbols to represent data points in graphs.

QUESTION 19

When conducting Earth and space science investigations, which of the following is a correct procedure for dealing with hazardous waste?

    A.   Pouring liquid waste down the sink with plenty of water
    B.   Mixing different types of waste together for easier disposal
    C.   Storing hazardous waste in labeled, leak-proof containers
    D.   Throwing hazardous waste in regular trash bins

QUESTION 20

Which of the following is an appropriate unit for measuring the volume of a liquid in Earth and space science?

    A.   Grams (g)
    B.   Meters (m)
    C.   Liters (L)
    D.   Seconds (s)

QUESTION 21

When using a balance to measure the mass of an object in Earth and space science investigations, which of the following should be done?

    A.   Place the object gently on the balance pan without taring it
    B.   Add weights to the object until the balance reads zero
    C.   Use the balance with wet hands to avoid electrostatic discharge
    D.   Read the mass from the position of the pointer on the balance scale

QUESTION 22

Which of the following is an appropriate use of scientific notation in reporting data in Earth and space science?

    A.   Describing qualitative observations of natural phenomena
    B.   Labeling the axes of a graph
    C.   Representing large distances such as the diameter of a planet
    D.   Writing the names of chemical compounds

Question 23

Which of the following is an appropriate safety measure when working with chemicals in the laboratory during Earth and space science investigations?

    A.   Smelling chemicals directly to identify them
    B.   Mixing chemicals randomly to observe reactions
    C.   Dispensing chemicals using the mouth
    D.   Labeling all chemical containers correctly

QUESTION 24

What is the appropriate unit for measuring the density of a substance in Earth and space science investigations?

    A.   Grams per liter (g/L)
    B.   Meters per second (m/s)
    C.   Newtons (N)
    D.   Degrees Celsius (°C)

QUESTION 25

When conducting experiments outdoors, what is an important safety consideration for Earth and space science teachers and students?

    A. Leaving equipment unattended to explore the surroundings
    B. Wearing sandals or open-toed shoes for comfort
    C. Ignoring weather forecasts as they may not be accurate
    D. Being aware of the location of emergency exits and shelters

QUESTION 26

What is the primary cause of Earth's seasons?

    A. Changes in the distance between the Earth and the Sun
    B. Variations in the Earth's axial tilt
    C. The eccentricity of the Earth's orbit
    D. The presence of the Moon's gravitational pull

QUESTION 27

Which type of rock is formed through the cooling and solidification of lava or magma?

    A. Sedimentary rock
    B. Metamorphic rock
    C. Igneous rock
    D. Tectonic rock

QUESTION 28

What is the process by which rocks are broken down into smaller particles through physical disintegration or chemical decomposition?

    A. Erosion
    B. Deposition
    C. Weathering
    D. Compaction

QUESTION 29

Which layer of the Earth's atmosphere contains the ozone layer?

    A. Troposphere
    B. Mesosphere
    C. Stratosphere
    D. Thermosphere

QUESTION 30

What type of plate boundary is associated with the formation of volcanoes and earthquakes?

    A. Transform boundary
    B. Convergent boundary
    C. Divergent boundary
    D. Subduction boundary

QUESTION 31

What scientific ideas contributed to the historical development of plate tectonics theory?

A. Ptolemaic geocentric model, luminiferous aether, and the phlogiston theory.
B. Geocentric model, phlogiston theory, and alchemy.
C. Continental drift, seafloor spreading, and subduction zones.
D. Ether theory, caloric theory, and the phlogiston theory.

QUESTION 32

Which major scientific idea has had a significant impact on our understanding of the history and development of the Universe?

A. Steady-state theory.
B. Geocentric model.
C. Heliocentric model.
D. Ptolemaic system.

QUESTION 33

Which major contemporary theory is fundamental to understanding the structure and behavior of matter?

A. Phlogiston theory.
B. Luminiferous aether theory.
C. Quantum mechanics.
D. Caloric theory.

QUESTION 34

Which unifying theme connects various branches of science and other STEM fields?

A. The theory of the four elements.
B. The theory of vitalism.
C. The theory of relativity.
D. The theory of evolution.

QUESTION 35

What is a fundamental characteristic of the nature of science as a system of inquiry?

A. Dogmatism.
B. Fixed methodology.
C. Subjectivity.
D. Empiricism.

QUESTION 36

Which major scientific concept is central to the understanding of the origin and evolution of species?

A. Germ theory.
B. Theory of special relativity.
C. Cell theory.
D. Theory of natural selection.

QUESTION 37

Which historical scientific idea significantly influenced the understanding of the Earth's geological processes and the concept of uniformitarianism?

A. Catastrophism.
B. Luminiferous aether theory.
C. Theory of vitalism.
D. Phlogiston theory.

QUESTION 38

Which major principle connects the fields of chemistry and biology and explains the continuity of life through genetic material?

A. Theory of the aether.
B. Theory of special relativity.
C. Theory of spontaneous generation.
D. Theory of genetics.

QUESTION 39

What is a key limitation of scientific models in the study of natural phenomena?

A. They are too detailed and comprehensive.
B. They are static and unchangeable.
C. They cannot predict future events with certainty.
D. They only apply to physics and not other sciences.

QUESTION 40

Which characteristic of the nature of science emphasizes the importance of revising scientific ideas based on new evidence?

A. Immutable principles.
B. Unchanging laws.
C. Tentativeness.
D. Dogmatic assertions.

QUESTION 41

Which major scientific concept explains the process by which rocks and minerals are broken down into smaller particles through physical, chemical, and biological processes?

A. Seafloor spreading.
B. Rock magnetism.
C. Plate tectonics.
D. Weathering.

QUESTION 42

Which geological feature is formed by the accumulation and compaction of organic material over millions of years?

A. Volcanic crater
B. Delta
C. Fossil fuel reservoir
D. Sinkhole

QUESTION 43

What is the primary source of energy that drives Earth's internal processes, including plate tectonics and volcanic activity?

A. Solar radiation.
B. Radioactive decay.
C. Gravitational forces.
D. Geothermal heat.

QUESTION 44

Which scientific principle describes the relationship between the velocity of a fluid (liquid or gas) and the pressure exerted by that fluid?

A. Boyle's law.
B. Archimedes' principle.
C. Bernoulli's principle.
D. Pascal's principle.

QUESTION 45

What major Earth science concept explains the formation, movement, and properties of glaciers?

A. Seafloor spreading.
B. Glacial isostasy.
C. Rock cycle.
D. Erosion.

QUESTION 46

Which historical figure is credited with proposing the heliocentric model of the solar system, placing the Sun at the center, and suggesting that Earth and other planets revolve around it?

A. Nicolaus Copernicus
B. Galileo Galilei
C. Johannes Kepler
D. Isaac Newton

QUESTION 47

What is the primary source of energy that drives Earth's climate and weather patterns?

A. Geothermal energy
B. Solar energy
C. Wind energy
D. Nuclear energy

QUESTION 48

The process of converting atmospheric nitrogen into a usable form for plants is known as:

A. Nitrogen fixation
B. Photosynthesis
C. Respiration
D. Transpiration

QUESTION 49

Which major scientific theory explains the movement of Earth's lithospheric plates and the formation of various landforms?

    A. Theory of Evolution
    B. Plate Tectonics
    C. Theory of Relativity
    D. Big Bang Theory

QUESTION 50

What type of rock is formed from the cooling and solidification of molten lava?

    A. Sedimentary rock
    B. Metamorphic rock
    C. Igneous rock
    D. Fossilized rock

QUESTION 51

The greenhouse effect is primarily caused by the accumulation of which gas in the Earth's atmosphere?

    A. Carbon dioxide (CO_2)
    B. Oxygen (O_2)
    C. Nitrogen (N_2)
    D. Argon (Ar)

QUESTION 52

Which branch of science studies the composition, structure, and properties of matter and the changes it undergoes during chemical reactions?

    A. Biology
    B. Physics
    C. Chemistry
    D. Geology

QUESTION 53

The scientific notation "10^7 years" would be most relevant in the context of:

    A. The age of Earth
    B. The duration of a solar day
    C. The period of a lunar cycle
    D. The time it takes for light to reach Earth from the Sun

QUESTION 54

Which scientific principle explains why an object submerged in a fluid experiences an upward buoyant force equal to the weight of the fluid displaced?

    A. Newton's Second Law
    B. Law of Conservation of Energy
    C. Archimedes' Principle
    D. Boyle's Law

QUESTION 55

In the scientific method, which step involves making predictions based on the proposed hypothesis?

A. Observation
B. Experimentation
C. Data Analysis
D. Formulating a hypothesis

QUESTION 56

Which major scientific theory explains the gradual change in Earth's magnetic field over time and the periodic reversal of magnetic poles?

A. Theory of Evolution
B. Plate Tectonics
C. Theory of Relativity
D. Theory of Paleomagnetism

QUESTION 57

Which type of rock is typically formed from the accumulation of mineral and organic particles at the bottom of lakes, seas, and oceans?

A. Igneous rock
B. Metamorphic rock
C. Sedimentary rock
D. Fossilized rock

QUESTION 58

Which process is responsible for the large-scale, long-term changes in Earth's climate due to variations in its orbit, axial tilt, and precession?

A. Greenhouse effect
B. El Niño-Southern Oscillation (ENSO)
C. Milankovitch cycles
D. Solar radiation

QUESTION 59

The Richter scale is used to measure the:

A. Intensity of hurricanes
B. Energy released by earthquakes
C. Wind speed of tornadoes
D. Pressure in volcanoes

QUESTION 60

Which type of volcano is characterized by a broad, flat shape with gentle slopes, and is primarily composed of basaltic lava flows?

A. Stratovolcano
B. Cinder cone volcano
C. Shield volcano
D. Caldera volcano

QUESTION 61

What is the main focus of Earth and space sciences?

    A.   Analyzing the interrelationships of science, technology, engineering, and mathematics
    B.   Evaluating scientific research and the validity of media coverage
    C.   Analyzing social, economic, and ethical issues associated with technological and scientific developments
    D.   Demonstrating knowledge of maps, models, and other geospatial technologies

Question 62

Which field of study explores the connections between science, technology, engineering, and mathematics in the context of Earth and space sciences?

    A.   Geospatial technology
    B.   Earth and space engineering
    C.   Earth and space mathematics
    D.   Earth and space science

QUESTION 63

What is the primary goal of evaluating scientific research and media coverage in Earth and space sciences?

    A.   To understand the technological and scientific developments
    B.   To analyze social, economic, and ethical issues
    C.   To ensure accurate representation of scientific information
    D.   To demonstrate knowledge of geospatial technologies

QUESTION 64

Which area of study focuses on exploring the social, economic, and ethical implications of scientific and technological developments?

    A.   Geospatial ethics
    B.   Earth and space economics
    C.   Technology and society
    D.   Earth and space engineering

QUESTION 65

What do maps, models, and geospatial technologies help earth science teachers with?

    A.   Analyzing scientific research and media coverage
    B.   Understanding social, economic, and ethical issues
    C.   Demonstrating the interrelationships of science, technology, engineering, and mathematics
    D.   Presenting scientific information effectively

QUESTION 66

In Earth and space sciences, what is the focus of analyzing social, economic, and ethical issues?

    A.   Examining the impact of science on technological advancements
    B.   Evaluating media coverage of scientific research
    C.   Understanding the interplay of science, technology, engineering, and mathematics
    D.   Investigating the implications of scientific and technological developments

QUESTION 67

Which of the following is NOT a part of the Earth and space sciences field?

A. Geology
B. Astronomy
C. Mathematics
D. Atmospheric science

QUESTION 68

What is the purpose of evaluating scientific research and media coverage in Earth and space sciences?

A. To explore geospatial technologies used in scientific information
B. To analyze the interrelationships between science, technology, engineering, and mathematics
C. To ensure accuracy and reliability in conveying scientific information
D. To understand the social and economic implications of scientific advancements

QUESTION 69

Which area of study involves understanding how scientific information is visually represented using maps, models, and geospatial technologies?

A. Earth and space communication
B. Geospatial science
C. Scientific visualization
D. Technology in earth sciences

QUESTION 70

What does the field of Earth and space sciences encompass?

A. The study of geospatial technologies and their impact on society
B. Analyzing media coverage and its influence on scientific research
C. The study of Earth, its processes, and the space environment
D. The examination of ethical issues related to scientific advancements

QUESTION 71

Which dating method helps geologists determine the age of rocks and fossils by comparing their positions in rock layers?

A. Radiometric dating
B. Absolute dating
C. Relative dating
D. Carbon dating

QUESTION 72

The geologic time scale is primarily based on:

A. Radioactive decay rates
B. Fossil record
C. Tree ring analysis
D. Ice core data

QUESTION 73

What type of dating technique uses the half-life of isotopes to determine the absolute age of rocks and minerals?

    A.   Carbon dating
    B.   Relative dating
    C.   Radiometric dating
    D.   Paleomagnetic dating

QUESTION 74

The major event that marks the end of the Mesozoic Era and the extinction of the dinosaurs is called:

    A.   The Cretaceous-Paleogene (K-Pg) boundary (Correct Answer)
    B.   The Permian-Triassic boundary
    C.   The Cambrian-Ordovician boundary
    D.   The Tertiary-Quaternary boundary

QUESTION 75

Which process contributed significantly to the formation of Earth's early atmosphere and hydrosphere?

    A.   Photosynthesis
    B.   Plate tectonics
    C.   Volcanic outgassing (Correct Answer)
    D.   Comet impacts

QUESTION 76

The first widely accepted scientific explanation for the origin of the solar system is the:

    A.   Nebular hypothesis
    B.   Panspermia theory
    C.   Steady-state theory
    D.   Big Bang theory

QUESTION 77

What process is responsible for the formation of fossils?

    A.   Erosion and weathering
    B.   Petrification
    C.   Volcanic eruptions
    D.   Meteor impacts

QUESTION 78

The fossil record provides evidence for:

    A.   The Big Bang theory
    B.   The theory of evolution
    C.   Plate tectonics
    D.   Radioactive decay

QUESTION 79

Which era is known as the "Age of Reptiles" due to the dominance of dinosaurs?

- A. Cenozoic Era
- B. Paleozoic Era
- C. Mesozoic Era (Correct Answer)
- D. Precambrian Era

QUESTION 80

The process by which a less complex organism evolves into a more complex one over time is known as:

- A. Natural selection
- B. Genetic mutation
- C. Adaptive radiation
- D. Biological evolution

QUESTION 81

The dominant gas in Earth's early atmosphere was:

- A. Oxygen
- B. Carbon dioxide
- C. Nitrogen
- D. Methane

QUESTION 82

Which dating technique uses the Earth's magnetic field to determine the age of rocks and minerals?

- A. Radiometric dating
- B. Paleomagnetic dating
- C. Relative dating
- D. Fossil dating

QUESTION 83

The theory of plate tectonics proposes that Earth's lithosphere is divided into:

- A. Three layers
- B. Five layers
- C. Seven layers
- D. Tectonic plates

QUESTION 84

The first organisms on Earth were likely:

- A. Birds
- B. Fish
- C. Prokaryotes (Correct Answer)
- D. Dinosaurs

QUESTION 85

The fossilized resin of ancient trees is known as:

A. Amber
B. Obsidian
C. Basalt
D. Lignite

QUESTION 86

The discovery of an unusual fossilized creature in sedimentary rocks from two different continents has sparked debate among geologists. Some argue that the fossil provides evidence for continental drift, while others remain skeptical. Explain the significance of this discovery and how it relates to the theory of continental drift. Which lines of evidence would you consider to support or refute the theory, and why?

A. The fossil discovery supports continental drift because it shows a similar species existing on two separated continents, providing evidence of past connections.
B. The fossil discovery does not relate to continental drift; it is likely an artifact of migration or a misinterpretation of geological processes.
C. The fossil discovery contradicts continental drift, as the species is unique and could not have traveled between the continents.
D. The fossil discovery requires more evidence to draw any conclusions about continental drift, as it may represent a new species or a common ancestor shared by both continents.

QUESTION 87

In a hypothetical scenario, a massive asteroid impact has been proposed as the cause of a mass extinction event. Evaluate the evidence and arguments that scientists might present to support this hypothesis. In your analysis, compare it to other potential extinction triggers, such as volcanic eruptions or climate change, and discuss the implications for understanding the history of life on Earth.

A. The asteroid impact hypothesis is the most plausible explanation, as it can produce the widespread devastation needed to cause a mass extinction.
B. The asteroid impact hypothesis is less likely than volcanic eruptions or climate change, as the effects of an impact are relatively localized.
C. The asteroid impact hypothesis has equal validity to volcanic eruptions or climate change, as all three events can lead to catastrophic consequences.
D. The asteroid impact hypothesis is supported by a range of evidence, such as the discovery of impact craters, shocked quartz, and iridium anomalies, making it a compelling explanation for the mass extinction event.

QUESTION 88

A group of researchers claims to have found the earliest known evidence of life on Earth, suggesting the existence of microbial mats in ancient rocks. Critically assess their findings and proposed interpretations. What alternative explanations could challenge their conclusions, and what further investigations or analyses could validate or refute their claims?

A. The researchers' findings are conclusive, as microbial mats are a well-known sign of early life. Further analyses of isotopic ratios could confirm their conclusions.
B. The researchers' findings are likely misinterpreted; the structures identified as microbial mats could be mineral formations or abiotic in origin.
C. The researchers' findings are promising, but more data are needed. Conducting DNA analysis on the structures would provide definitive evidence of ancient life.
D. The researchers' findings are intriguing, but microbial mats can also form in modern environments, making it challenging to determine their ancient origin. Conducting electron microscopy could confirm their biological nature.

QUESTION 89

The fossil record indicates that certain species experienced rapid and remarkable diversification during specific periods in Earth's history. Explain the concept of adaptive radiation and how it contributed to the evolution of diverse species. Provide examples of well-known instances of adaptive radiation and describe the environmental factors that likely facilitated these events.

A. Adaptive radiation occurs when species become extinct, allowing for new species to fill available niches, leading to remarkable diversity. An example is the Cretaceous-Tertiary extinction event, which led to the diversification of mammals after the extinction of dinosaurs.

B. Adaptive radiation is the process of species adapting to new environments, leading to speciation. An example is the Galapagos finches, which evolved different beak shapes and sizes due to varying food sources on different islands.

C. Adaptive radiation is the result of multiple species converging to form a single, highly specialized species. An example is the evolution of bats from various flying insects.

D. Adaptive radiation occurs when a single species spreads out to occupy different habitats, leading to the formation of multiple new species. An example is the colonization of different island groups by lemurs, resulting in the evolution of various lemur species.

QUESTION 90

The discovery of transitional fossils, such as Archaeopteryx, has been pivotal in supporting the theory of evolution. Analyze the significance of Archaeopteryx and other transitional fossils in explaining the evolutionary relationship between dinosaurs and birds. What anatomical features and characteristics of these fossils provide compelling evidence for the evolution of birds from dinosaurs?

A. Archaeopteryx and other transitional fossils show the direct transformation of dinosaurs into birds, proving the accuracy of the theory of evolution.

B. Archaeopteryx and other transitional fossils exhibit a mix of dinosaur and bird characteristics, suggesting an evolutionary link between the two groups.

C. Archaeopteryx and other transitional fossils are likely a separate group of organisms with no evolutionary connection to dinosaurs or birds.

D. Archaeopteryx and other transitional fossils provide evidence for parallel evolution, with dinosaurs and birds developing similar features independently.

QUESTION 91

What is the primary driving force behind tectonic plate movement?

A. Gravity from the Earth's core.
B. Convection currents in the mantle.
C. The pull of neighboring celestial bodies.
D. Friction between the plates.

QUESTION 92

Which type of volcano is characterized by highly explosive eruptions and steep-sided slopes due to the buildup of viscous lava?

A. Shield volcano.
B. Cinder cone volcano.
C. Stratovolcano (or composite volcano).
D. Lava dome volcano.

QUESTION 93

How are earthquakes caused along tectonic plate boundaries?

A. The collision of two tectonic plates.
B. The sliding of tectonic plates past each other.
C. The separation of tectonic plates.
D. The convergence of tectonic plates with a subduction zone.

QUESTION 94

Which process involves the transformation of existing rock into a new rock through changes in temperature and pressure within the Earth's crust?

    A.   Weathering and erosion
    B.   Metamorphism
    C.   Melting and solidification
    D.   Deposition and lithification

QUESTION 95

Which type of igneous rock is characterized by rapid cooling at or near the Earth's surface, resulting in fine-grained texture?

    A.   Granite
    B.   Gabbro
    C.   Basalt
    D.   Pumice

QUESTION 96

What type of rock is formed from the accumulation and cementation of mineral and organic particles in various environments?

    A.   Metamorphic rock.
    B.   Igneous rock.
    C.   Sedimentary rock.
    D.   Conglomerate rock.

QUESTION 97

Which type of plate boundary is associated with the formation of mid-ocean ridges?

    A.   Divergent boundary.
    B.   Convergent boundary.
    C.   Transform boundary.
    D.   Subduction boundary.

QUESTION 98

What type of evidence is used to support the theory of plate tectonics?

    A.   Fossil records.
    B.   Paleomagnetic data.
    C.   Earthquake intensity maps.
    D.   Meteorite impacts.

QUESTION 99

Which type of volcanic feature is a large, steep-walled depression that forms following a massive eruption, often due to the collapse of the magma chamber?

    A.   Caldera.
    B.   Lava tube.
    C.   Maar.
    D.   Tephra cone.

QUESTION 100

What process is responsible for the gradual wearing away of Earth's surface through the action of wind, water, ice, and gravity?

A. Deposition.
B. Erosion.
C. Subduction.
D. Volcanism.

QUESTION 101

Which type of fault is characterized by the movement of rock blocks away from each other due to tensional forces?

A. Normal fault.
B. Reverse fault.
C. Strike-slip fault.
D. Thrust fault.

QUESTION 102

The theory of plate tectonics has significantly transformed our understanding of the Earth's geology and processes. However, there are still some misconceptions and challenges in its acceptance. Discuss one of the main misconceptions or challenges related to the theory of plate tectonics, and provide an explanation for why it persists.

A. The challenge of reconciling the theory with catastrophic events like meteorite impacts.
B. The misconception that plate tectonics is primarily driven by the Earth's rotation.
C. The challenge of explaining the distribution of ancient fossils across continents.
D. The misconception that plate tectonics only affects landmasses and not oceanic regions.

QUESTION 103

To address this challenge, paleontologists and geologists propose several explanations. One possibility is that the organisms were able to disperse over temporary land bridges or island chains that existed in the past but have since disappeared due to plate tectonic movements. Another explanation is that some fossils may have been transported by ocean currents, storms, or floating debris, allowing for limited connectivity between continents. The Earth's rock cycle plays a crucial role in the formation and transformation of different rock types. Discuss one specific example of how the rock cycle operates in nature, involving all three major rock types (igneous, sedimentary, and metamorphic), and describe the processes involved.

A. The formation of a sedimentary rock through weathering and erosion, followed by metamorphism, and subsequent melting to produce an igneous rock.
B. The transformation of an igneous rock into a sedimentary rock through compaction and cementation, followed by melting and solidification into a new igneous rock.
C. The formation of an igneous rock through volcanic activity, followed by erosion and transportation to produce sedimentary rock, and further metamorphism into a metamorphic rock.
D. The transformation of a metamorphic rock into an igneous rock through melting, followed by weathering and erosion to form sedimentary rock.

QUESTION 104

Earthquakes can have devastating impacts on both human communities and the environment. As an earth science teacher, you are tasked with educating your students on earthquake preparedness and mitigation strategies. Outline three essential earthquake preparedness measures that individuals and communities can implement, and explain the rationale behind each measure.

    A. Constructing buildings with flexible materials, creating emergency response plans, and securing heavy furniture.
    B. Building earthquake-resistant structures, stockpiling food and water, and avoiding construction in seismically active zones.
    C. Developing early warning systems, practicing "Drop, Cover, and Hold On" drills, and reinforcing critical infrastructure.
    D. Promoting public awareness campaigns, implementing earthquake insurance, and conducting regular seismic risk assessments.

QUESTION 105

The Earth's geologic history is marked by significant mass extinctions that have profoundly impacted the diversity of life on the planet. Describe one of the major mass extinctions in Earth's history, and discuss the prevailing theories or hypotheses regarding its causes.

    A. The Cretaceous-Paleogene (K-Pg) mass extinction, possibly caused by volcanic activity or asteroid impact.
    B. The Permian-Triassic (P-T) mass extinction, potentially linked to widespread glaciation and sea-level changes.
    C. The Ordovician-Silurian mass extinction, likely triggered by large-scale volcanic eruptions.
    D. The Triassic-Jurassic mass extinction, possibly influenced by an increase in greenhouse gases and global warming.

QUESTION 106

Which of the following statements about minerals is correct?

    A. Minerals are man-made substances with a specific crystal structure.
    B. Minerals can be composed of organic materials only.
    C. Quartz and feldspar are examples of minerals.
    D. Minerals are always formed through biological processes.

QUESTION 107

What is the process of extracting useful minerals or rocks from the Earth's crust called?

    A. Bioremediation
    B. Mining
    C. Geothermal energy extraction
    D. Fossilization

QUESTION 108

Which type of rock is formed from the cooling and solidification of magma or lava?

    A. Sedimentary rock
    B. Metamorphic rock
    C. Igneous rock
    D. Tectonic rock

QUESTION 109

What is the study of seismic waves and their behavior as they pass through different layers of the Earth called?

    A. Geodesy
    B. Seismology
    C. Petrology
    D. Paleontology

QUESTION 110

Which of the following is NOT a piece of evidence for the Earth's internal structure?

A. Seismic wave behavior
B. Magnetic field measurements
C. Fossil records
D. Rock samples from drilling

QUESTION 111

Which process involves the physical and chemical breakdown of rocks into smaller particles due to natural processes like water, wind, and temperature changes?

A. Weathering
B. Metamorphism
C. Subduction
D. Crystallization

QUESTION 112

What type of soil is characterized by having roughly equal proportions of sand, silt, and clay?

A. Loam
B. Peat
C. Gravel
D. Chalk

QUESTION 113

Which of the following is a non-renewable geologic resource?

A. Wind energy
B. Solar energy
C. Natural gas
D. Biomass

QUESTION 114

Which layer of the Earth's interior is primarily composed of molten iron and nickel?

A. Crust
B. Mantle
C. Outer core
D. Inner core

QUESTION 115

What is the process called by which one tectonic plate moves under another tectonic plate and sinks into the Earth's mantle?

A. Subduction
B. Seafloor spreading
C. Rifting
D. Faulting

QUESTION 116

What type of rock forms from the alteration of existing rock through heat, pressure, or chemically active fluids?

    A. Sedimentary rock
    B. Igneous rock
    C. Metamorphic rock
    D. Mineral rock

QUESTION 117

Which of the following minerals is the hardest known naturally occurring substance on Earth?

    A. Quartz
    B. Diamond
    C. Calcite
    D. Gypsum

QUESTION 118

Which of the following soil horizons contains organic matter in various stages of decomposition?

    A. A horizon
    B. B horizon
    C. C horizon
    D. A horizon

QUESTION 119

What geologic resource is formed from the remains of plants and animals that lived millions of years ago and has been subjected to heat and pressure?

    A. Coal
    B. Iron ore
    C. Limestone
    D. Gypsum

QUESTION 120

What method is commonly used to determine the internal structure of the Earth by analyzing how seismic waves travel through the planet?

    A. Core sampling
    B. Drilling
    C. Seismology
    D. Volcanic eruption analysis

QUESTION 121

Volcanic eruptions have significant impacts on the environment and communities. As an Earth Science teacher, you want to discuss the potential consequences of a major volcanic eruption in a densely populated area. Which of the following statements about the impacts of a volcanic eruption is correct?

    A. The primary concern during a volcanic eruption is the release of harmful radioactive materials into the atmosphere.
    B. Volcanic eruptions can lead to the formation of new minerals that are beneficial for agriculture and construction.
    C. The eruption's immediate effects include the release of toxic gases, ash, and pyroclastic flows, which can pose serious hazards to people and the environment.
    D. Volcanic eruptions have no long-term effects on the climate and weather patterns.

QUESTION 122

Plate tectonics play a crucial role in shaping the Earth's surface and geological processes. As a teacher, you want to explain the concept of plate tectonics to your students. Which of the following statements best describes plate tectonics?

A. Plate tectonics refer to the study of ancient fossils found on separate continents, which provide evidence for continental drift.
B. Plate tectonics is the process of volcanic eruptions that lead to the formation of new ocean basins.
C. Plate tectonics is the theory that Earth's lithosphere is divided into several large and small plates that move and interact with each other on the semi-fluid asthenosphere.
D. Plate tectonics primarily deals with the study of meteorite impacts on Earth's surface and their effects on geological formations.

QUESTION 123

As an Earth Science teacher, you want to discuss the significance of soil in supporting life on Earth. Which of the following statements about soil is accurate?

A. Soil is a renewable resource that can be rapidly replenished through natural processes such as erosion and weathering.
B. Soil primarily serves as a medium for holding water, and its nutrient content does not play a significant role in supporting plant growth.
C. Soil is a complex mixture of mineral particles, organic matter, water, air, and living organisms, making it a critical component for supporting plant growth and providing habitats for various organisms.
D. Soil formation is a straightforward process that does not require the interaction of multiple factors over long periods of time.

QUESTION 124

One of your students asks you about the difference between a rock and a mineral. Provide an accurate explanation to help them understand the distinction. Which of the following statements best clarifies the difference between rocks and minerals?

A. Minerals are formed through the cooling and solidification of molten rock, while rocks are naturally occurring, inorganic substances with a definite chemical composition.
B. Rocks are primarily composed of a single type of mineral, while minerals are composed of a mixture of different rocks.
C. Minerals are only found on the Earth's surface, while rocks are found deep within the Earth's interior.
D. Rocks are the result of biological processes, while minerals are inorganic substances formed through geological processes.

QUESTION 125

Climate change and its impact on Earth's systems have become a significant concern for our planet's future. As an Earth Science teacher, you want to highlight the relationship between human activities, greenhouse gas emissions, and global warming. Which of the following statements about climate change is accurate?

A. Human activities have no significant impact on climate change, as natural processes are solely responsible for the increase in greenhouse gas concentrations.
B. Greenhouse gases primarily consist of oxygen and nitrogen, which are released into the atmosphere through volcanic eruptions and geothermal activities.
C. Burning fossil fuels, deforestation, and industrial processes are some of the human activities that release greenhouse gases into the atmosphere, leading to an enhanced greenhouse effect and global warming.
D. Global warming is a natural cycle that occurs over millions of years and has no link to human activities or greenhouse gas emissions.

QUESTION 126

Which geological principle states that in a sequence of undisturbed layers of rock, the oldest rocks are at the bottom and the youngest rocks are at the top?

A. Law of Superposition
B. Law of Original Horizontality
C. Law of Cross-Cutting Relationships
D. Law of Faunal Succession

QUESTION 127

What type of fossil is most useful for correlating and dating rock layers over large distances?

- A. Index fossil
- B. Trace fossil
- C. Cast fossil
- D. Mold fossil

QUESTION 128

The process by which an unstable atomic nucleus loses energy by emitting radiation is called:

- A. Radioactive decay
- B. Nuclear fission
- C. Nuclear fusion
- D. Half-life

QUESTION 129

Which type of plate boundary is characterized by two plates moving away from each other?

- A. Convergent boundary
- B. Divergent boundary
- C. Transform boundary
- D. Subduction boundary

QUESTION 130

The process by which one tectonic plate moves beneath another into the mantle is known as:

- A. Subduction
- B. Rifting
- C. Folding
- D. Faulting

QUESTION 131

What type of rock forms from the cooling and solidification of magma or lava?

- A. Sedimentary rock
- B. Metamorphic rock
- C. Igneous rock
- D. Rock fragment

QUESTION 132

Which of the following is NOT one of the main compositional layers of the Earth?

- A. Crust
- B. Mantle
- C. Core
- D. Lithosphere

QUESTION 133

What type of mineral is a naturally occurring, inorganic solid with a specific chemical composition and a regular internal crystal structure?

A. Rock
B. Gem
C. Metal
D. Mineral

QUESTION 134

The sustainable and responsible use of Earth's materials and resources to meet current and future needs is known as:

A. Geologic engineering
B. Resource management
C. Mineral exploitation
D. Environmental geology

QUESTION 135

Which type of weathering is caused by the physical disintegration of rock without any chemical change in the minerals?

A. Chemical weathering
B. Biological weathering
C. Mechanical weathering
D. Oxidative weathering

QUESTION 136

The process of sediment being laid down in new locations is called:

A. Erosion
B. Weathering
C. Deposition
D. Transportation

QUESTION 137

Which erosional landform is characterized by an amphitheater-like basin with steep walls, often formed by glacial erosion?

A. Meander
B. Delta
C. Cirque
D. Alluvial fan

QUESTION 138

During a field trip, students discover two sedimentary rock layers separated by an unconformity. The bottom layer contains shale with abundant marine fossils, and the upper layer contains sandstone with terrestrial plant fossils. How would you interpret this sequence of rocks and the unconformity between them? Choose the best option:

A. The unconformity represents a period of non-deposition or erosion, followed by the deposition of the sandstone layer. The sequence provides evidence of a marine environment transitioning to a terrestrial environment over time.
B. The unconformity is an example of a disconformity, indicating a period of non-deposition. The shale layer was uplifted and exposed to erosion before the sandstone layer was deposited.
C. The rocks show evidence of being originally deposited horizontally, but tectonic forces caused them to fold and tilt. The shale layer was uplifted and eroded before the sandstone layer was deposited.
D. The presence of marine and terrestrial fossils indicates that the unconformity is a nonconformity, suggesting the sandstone layer was uplifted, eroded, and then covered by marine sediments.

QUESTION 139

Imagine two volcanic islands situated on the ocean floor, several hundred kilometers apart. Both islands have the same type of volcanic rocks and similar marine fossils in their surrounding sediments. However, the ages of the volcanic rocks on Island A are much older than those on Island B. How could you explain this geological phenomenon? Choose the best option:

A. Island A and Island B are part of the same tectonic plate, but Island A is located farther away from the mid-ocean ridge. The rocks on Island A cooled and solidified earlier than those on Island B.
B. The volcanic rocks and fossils on both islands are unrelated. Island A experienced multiple volcanic eruptions over a long geological history, while Island B experienced a more recent and shorter volcanic activity.
C. Island A and Island B were once part of the same landmass. Due to plate tectonics, they drifted apart, and the volcanic rocks on Island A became older through the process of isostatic subsidence.
D. The volcanic rocks on Island A were formed by subduction-related volcanism, whereas Island B experienced hotspot volcanism. The differences in tectonic settings account for the age disparity between the rocks on the two islands.

QUESTION 140

A mining company claims to have discovered a vast deposit of a valuable mineral in a remote area. The company presents core samples showing the presence of the mineral in the subsurface. As an Earth Science teacher, what critical factors would you consider to assess the economic viability and potential environmental impacts of the mining project? Choose the best option:

A. Analyze the mineral's physical properties and chemical composition to determine its market demand and economic value. Evaluate the accessibility of the site, potential transportation costs, and the environmental impact of mining activities on the surrounding ecosystem and local communities.
B. Estimate the total volume of the mineral deposit and its purity to calculate the potential yield and revenue. Conduct a social impact assessment to understand the local community's perspectives and potential conflicts with indigenous populations.
C. Investigate the historical market trends of the mineral and its global demand. Evaluate the feasibility of setting up mining infrastructure in the remote area and develop a comprehensive reclamation plan to restore the site after mining operations.
D. Assess the geotechnical aspects of the deposit, such as its depth, stability, and potential risks of geological hazards. Calculate the cost of extracting the mineral and evaluate its potential contribution to the company's profits.

QUESTION 141

What is the primary difference between physical and chemical weathering?

A. Physical weathering involves the breakdown of rocks without any chemical changes, while chemical weathering involves the alteration of rock composition through chemical reactions.
B. Physical weathering only occurs in hot and dry climates, while chemical weathering occurs in cold and wet climates.
C. Physical weathering is a slow process that takes thousands of years, while chemical weathering is a rapid process that occurs within a few weeks.
D. Physical weathering occurs primarily in aquatic environments, while chemical weathering occurs on land.

QUESTION 142

Which erosional process is responsible for the formation of U-shaped valleys in mountainous regions?

A. Abrasion by wind
B. Wave erosion along coastlines
C. Glacial erosion by alpine glaciers
D. Soil erosion due to deforestation

QUESTION 143

Which of the following processes is responsible for sorting sediments by size as they are transported by water?

A. Erosion
B. Weathering
C. Deposition
D. Sedimentation

QUESTION 144

What landforms are created by the process of chemical weathering?

    A.   Caves and sinkholes
    B.   Moraines and drumlins
    C.   U-shaped valleys and hanging valleys
    D.   Deltas and alluvial fans

QUESTION 145

Which of the following is NOT an erosional process?

    A.   Wind erosion
    B.   Glacier erosion
    C.   River erosion
    D.   Volcanic eruption

QUESTION 146

Which glacier is responsible for shaping fjords along coastal areas?

    A.   Alpine glaciers
    B.   Continental glaciers
    C.   Ice shelves
    D.   Piedmont glaciers

QUESTION 147

How does the size of sediment particles influence their transport by water?

    A.   Larger particles are carried further due to their greater buoyancy.
    B.   Smaller particles settle quickly and are transported shorter distances.
    C.   Larger particles settle more slowly and are deposited closer to the source.
    D.   Smaller particles are transported faster due to their higher density.

QUESTION 148

Which factor primarily determines the rate of chemical weathering in an area?

    A.   Temperature and humidity
    B.   Elevation and precipitation
    C.   Wind speed and direction
    D.   Earthquake frequency

QUESTION 149

What is the main difference between weathering and erosion?

    A.   Weathering refers to the breakdown of rocks, while erosion involves the movement and transportation of the resulting sediments.
    B.   Weathering occurs in aquatic environments, while erosion occurs on land.
    C.   Weathering is primarily caused by volcanic activity, while erosion is driven by glaciation.
    D.   Weathering is a rapid process, while erosion is a slow process that takes thousands of years.

QUESTION 150

What type of glacier forms when a valley glacier flows out of the mountains and spreads into a broad lobe-like shape?

A. Hanging glacier
B. Piedmont glacier
C. Tidewater glacier
D. Cirque glacier

QUESTION 151

How does deforestation impact the process of erosion in terrestrial environments?

A. Deforestation reduces erosion by increasing the stability of the soil.
B. Deforestation has no impact on erosion rates in terrestrial environments.
C. Deforestation accelerates erosion due to the removal of vegetation that holds soil in place.
D. Deforestation reduces erosion by promoting the growth of deep-rooted plants.

QUESTION 152

What type of erosion is responsible for the formation of sand dunes in deserts?

A. Glacial erosion
B. Wind erosion
C. River erosion
D. Coastal erosion

QUESTION 153

Which factor influences the rate of glacial erosion?

A. Latitude of the glacier's location
B. Age of the glacier
C. Size of the glacier
D. Presence of vegetation near the glacier

QUESTION 154

How do variations in climate and geography affect the landscape?

A. Climate and geography have no impact on the landscape.
B. Changes in climate and geography can lead to shifts in the Earth's magnetic field.
C. Variations in climate and geography can result in the formation of different landforms, such as deserts, mountains, and river valleys.
D. Climate and geography primarily affect the atmosphere and have little influence on the landscape.

QUESTION 155

What is the correct order of the hydrologic cycle stages?

A. Evaporation, Condensation, Precipitation, Infiltration
B. Precipitation, Evaporation, Infiltration, Condensation
C. Condensation, Evaporation, Precipitation, Infiltration
D. Infiltration, Condensation, Evaporation, Precipitation

QUESTION 156

Which of the following energy changes primarily drives the water cycle?

    A.   Kinetic energy
    B.   Gravitational potential energy
    C.   Thermal energy
    D.   Nuclear energy

QUESTION 157

What process is responsible for converting water vapor into clouds?

    A.   Evaporation
    B.   Condensation
    C.   Precipitation
    D.   Infiltration

QUESTION 158

Which of the following is a chemical property of water?

    A.   Boiling point
    B.   Color
    C.   Taste
    D.   pH level

QUESTION 159

How does water chemistry change during the hydrologic cycle?

    A.   The pH level decreases during evaporation and increases during condensation.
    B.   The salinity increases during precipitation and decreases during infiltration.
    C.   The dissolved oxygen content increases during evaporation and decreases during condensation.
    D.   The nutrient concentration increases during precipitation and decreases during infiltration.

QUESTION 160

Which Earth system is NOT directly interconnected with the hydrosphere?

    A.   Biosphere
    B.   Atmosphere
    C.   Lithosphere
    D.   Cryosphere

QUESTION 161

How do ocean currents impact the Earth's climate?

    A.   They have a minimal effect on climate patterns.
    B.   They distribute heat around the Earth, influencing regional climates.
    C.   They only affect coastal regions and not global climate.
    D.   They are driven solely by atmospheric circulation and do not impact climate.

QUESTION 162

What is the primary source of water vapor in the atmosphere?

A. Rivers and lakes
B. Melting glaciers
C. Oceans and seas
D. Groundwater discharge

QUESTION 163

Which of the following is an example of a point source of water pollution?

A. Agricultural runoff
B. Industrial discharge
C. Urban stormwater runoff
D. Natural erosion

QUESTION 164

What is the significance of wetlands in the context of the hydrosphere and other Earth systems?

A. Wetlands act as reservoirs for freshwater storage.
B. They serve as barriers to prevent coastal erosion.
C. Wetlands have no substantial impact on Earth systems.
D. They solely support aquatic biodiversity and have no influence on other systems.

QUESTION 165

Which process involves the conversion of a solid directly to water vapor without becoming a liquid first?

A. Sublimation
B. Condensation
C. Precipitation
D. Infiltration

QUESTION 166

How do human activities contribute to changes in the chemical properties of water during the hydrologic cycle?

A. By increasing atmospheric carbon dioxide concentration.
B. By reducing the usage of fertilizers in agriculture.
C. By decreasing industrial discharge of pollutants.
D. By promoting reforestation and afforestation efforts.

QUESTION 167

What is the term for the process by which water in plants is lost to the atmosphere?

A. Transpiration
B. Infiltration
C. Evaporation
D. Percolation

QUESTION 168

What is the primary driving force of the hydrologic cycle?

    A.   Gravity
    B.   Solar radiation
    C.   Wind patterns
    D.   Earth's magnetic field

QUESTION 169

Which ocean layer contains the highest concentration of dissolved oxygen?

    A.   Epipelagic zone
    B.   Mesopelagic zone
    C.   Bathypelagic zone
    D.   Abyssopelagic zone

QUESTION 170

Which of the following is NOT a type of freshwater system?

    A.   River
    B.   Lake
    C.   Estuary
    D.   Fjord

QUESTION 171

What is the term for the process of water vapor turning directly into ice without passing through the liquid state?

    A.   Sublimation
    B.   Condensation
    C.   Precipitation
    D.   Melting

QUESTION 172

The Gulf Stream is an example of which oceanic feature?

    A.   Ocean trench
    B.   Ocean ridge
    C.   Ocean current
    D.   Ocean upwelling

QUESTION 173

Which process involves the movement of water through the soil and underlying rock layers?

    A.   Evaporation
    B.   Transpiration
    C.   Infiltration
    D.   Percolation

QUESTION 174

Which factor is NOT responsible for causing ocean tides?

    A.   Gravitational pull of the Moon
    B.   Gravitational pull of the Sun
    C.   Earth's magnetic field
    D.   Centrifugal force

QUESTION 175

Which of the following is a characteristic feature of a braided river system?

    A.   Straight and narrow channel
    B.   Single, meandering channel
    C.   Multiple interconnected channels
    D.   Steep and deep valley

QUESTION 176

What is the major source of dissolved salts in seawater?

    A.   River runoff
    B.   Volcanic eruptions
    C.   Marine organisms
    D.   Atmospheric deposition

QUESTION 177

Which freshwater ecosystem is characterized by slow-moving or stagnant water and is often covered with a layer of floating vegetation?

    A.   Lake
    B.   Pond
    C.   Wetland
    D.   River

QUESTION 178

What is the term for the process of water vapor turning into liquid water as it cools?

    A.   Evaporation
    B.   Condensation
    C.   Sublimation
    D.   Precipitation

QUESTION 179

The Coriolis effect influences the direction of ocean currents due to:

    A.   Differences in water salinity
    B.   Differences in water temperature
    C.   Earth's rotation
    D.   Earth's magnetic field

QUESTION 180

Which process involves the loss of water vapor from plants through their leaves?

    A.   Infiltration
    B.   Percolation
    C.   Transpiration
    D.   Evaporation

QUESTION 181

What is the term for the area where groundwater fills all the available spaces in sediment or rock layers?

    A.   Water table
    B.   Aquifer
    C.   Aquitard
    D.   Porous zone

QUESTION 182

Which factor primarily determines the salinity levels of different oceans?

    A.   Latitude
    B.   Depth
    C.   Distance from the coastline
    D.   Ocean currents

QUESTION 183

What is the primary factor affecting stream flow in a watershed?

    A.   Temperature variations
    B.   Human activities
    C.   Precipitation patterns
    D.   Geological formations

QUESTION 184

Which of the following is a characteristic property of surface water?

    A.   High mineral content
    B.   Slow response to weather changes
    C.   Easily accessible for direct human use
    D.   Uniform temperature throughout the year

QUESTION 185

What factor affects the movement of groundwater in an aquifer?

    A.   Surface vegetation
    B.   Surface water flow
    C.   Wind speed
    D.   Permeability of the aquifer

QUESTION 186

Which geological factor plays a crucial role in determining the availability of freshwater resources?

    A. Presence of fossils
    B. Rock color
    C. Geological age of rock formations
    D. Aquifer thickness

QUESTION 187

What term refers to the area of land where all water drains to a single point, such as a river or lake?

    A. Watershed
    B. Aquifer
    C. Alluvium
    D. Groundwater recharge zone

QUESTION 188

Which of the following activities is most likely to affect the quality of surface water negatively?

    A. Constructing rainwater harvesting systems
    B. Implementing wastewater treatment plants
    C. Installing water-efficient irrigation systems
    D. Discharging untreated industrial waste into rivers

QUESTION 189

What is the primary mode of water movement in a confined aquifer?

    A. Horizontal flow
    B. Vertical flow
    C. Lateral flow
    D. Radial flow

QUESTION 190

Which of the following factors affects groundwater infiltration?

    A. Solar radiation
    B. Soil composition
    C. Stream flow rate
    D. Temperature fluctuations

QUESTION 191

What type of drainage pattern resembles the branches of a tree, with tributaries joining a main river or stream?

    A. Radial pattern
    B. Dendritic pattern
    C. Trellis pattern
    D. Centripetal pattern

QUESTION 192

Which of the following factors is most likely to affect the discharge of a river?

- A. Depth of the river
- B. Width of the river
- C. Velocity of the river
- D. Shape of the riverbed

QUESTION 193

Which geological factor affects the storage capacity of an aquifer?

- A. Rock color
- B. Geological age of rock formations
- C. Presence of fossils
- D. Porosity of the rock

QUESTION 194

What is the term for the process of water passing from the atmosphere to the Earth's surface as rain or snow?

- A. Evaporation
- B. Precipitation
- C. Condensation
- D. Transpiration

QUESTION 195

Which of the following is a common human activity that can lead to the depletion of groundwater resources?

- A. Installing rainwater harvesting systems
- B. Reducing water consumption
- C. Over-pumping from wells
- D. Recharging aquifers using treated wastewater

QUESTION 196

What is the term for the movement of water across the land surface, often due to heavy rainfall or rapid snowmelt?

- A. Infiltration
- B. Runoff
- C. Percolation
- D. Evaporation

QUESTION 197

Which of the following is NOT a factor that affects the quality of groundwater resources?

- A. Agricultural runoff
- B. Industrial discharge
- C. Surface water flow rate
- D. Natural mineral leaching

QUESTION 198

What is the primary factor driving ocean currents?

    A.   Gravitational pull from the Moon and the Sun
    B.   Earth's magnetic field
    C.   Wind patterns
    D.   Tidal waves

QUESTION 199

Which ocean zone receives the most sunlight and supports the highest biodiversity?

    A.   Abyssal zone
    B.   Bathyal zone
    C.   Epipelagic zone
    D.   Hadal zone

QUESTION 200

Which process is responsible for the formation of an oxbow lake?

    A.   Erosion
    B.   Deposition
    C.   Volcanic activity
    D.   Earthquake

QUESTION 201

Which of the following is an example of a freshwater wetland?

    A.   Coral reef
    B.   Estuary
    C.   Mangrove forest
    D.   Marsh

QUESTION 202

What causes tides on Earth?

    A.   Solar flares
    B.   Gravitational pull from other planets
    C.   Gravitational interaction between Earth, Moon, and Sun
    D.   Ocean currents

QUESTION 203

Which oceanic feature is a result of the divergence of tectonic plates?

    A.   Trench
    B.   Seamount
    C.   Mid-ocean ridge
    D.   Abyssal plain

QUESTION 204

The salinity of seawater is primarily influenced by:

- A. River runoff
- B. Evaporation
- C. Tidal activity
- D. Iceberg melting

QUESTION 205

What is the main factor responsible for coastal erosion?

- A. Human construction
- B. Storm surges
- C. Sea level rise
- D. Beach nourishment

QUESTION 206

Which ocean is the largest and deepest on Earth?

- A. Atlantic Ocean
- B. Indian Ocean
- C. Arctic Ocean
- D. Pacific Ocean

QUESTION 207

What is the primary cause of coral bleaching in coral reefs?

- A. Pollution from industrial runoff
- B. Overfishing
- C. Ocean acidification
- D. Elevated sea temperatures

QUESTION 208

Which freshwater ecosystem is characterized by low oxygen levels and decaying organic matter?

- A. Lake
- D. River
- C. Pond
- D. Bog

QUESTION 209

Which natural event is responsible for the formation of barrier islands along coastlines?

- A. Tsunamis
- B. Volcanic eruptions
- C. Storm surges
- D. Glacial melting

QUESTION 210

The Gulf Stream is an example of a:

    A.   Warm ocean current
    B.   Cold ocean current
    C.   Deep ocean current
    D.   Subsurface ocean current

QUESTION 211

Which of the following is a major source of freshwater for human use?

    A.   Glaciers and ice caps
    B.   Oceans
    C.   Deep underground aquifers
    D.   Saltwater lakes

QUESTION 212

What is the primary factor influencing the salinity of a freshwater lake?

    A.   Sunlight exposure
    B.   Precipitation patterns
    C.   Depth of the lake
    D.   Surrounding vegetation

QUESTION 213

What is the primary factor influencing the origin of ocean basins?

    A.   Plate tectonics
    B.   Solar radiation
    C.   Wind patterns
    D.   Earth's magnetic field

QUESTION 214

Which of the following coastal features is formed due to longshore drift?

    A.   Barrier islands
    B.   Sea arches
    C.   Fiords
    D.   Abyssal plains

QUESTION 215

What is the average salinity of ocean water worldwide?

    A.   3.5%
    B.   2.0%
    C.   1.0%
    D.   0.5%

QUESTION 216

Which ocean current is responsible for the warm waters along the eastern coast of the United States?

    A. Gulf Stream
    B. California Current
    C. Canary Current
    D. Labrador Current

QUESTION 217

What causes the formation of tides on Earth?

    A. Gravitational pull of the Moon and the Sun
    B. Wind patterns
    C. Earth's rotation
    D. Ocean currents

QUESTION 218

Which of the following marine resources is considered a non-living (abiotic) resource?

    A. Fish
    B. Coral reefs
    C. Oil deposits
    D. Algae

QUESTION 219

What type of coastline features numerous winding inlets, often resembling the letter "W"?

    A. Ria coastline
    B. Delta coastline
    C. Emergent coastline
    D. Fjord coastline

QUESTION 220

Which oceanic zone receives the least amount of sunlight and is home to bioluminescent organisms?

    A. Epipelagic zone
    B. Mesopelagic zone
    C. Bathypelagic zone
    D. Abyssopelagic zone

QUESTION 221

Which ocean current is responsible for bringing cold waters from the Arctic southward along the eastern coast of Canada and the United States?

    A. Gulf Stream
    B. California Current
    C. Canary Current
    D. Labrador Current

QUESTION 222

What are the biologically productive regions of the ocean where nutrient upwelling supports a high abundance of marine life?

A. Dead zones
B. Coral reefs
C. Estuaries
D. Upwelling zones

QUESTION 223

Which oceanic feature is characterized by underwater mountain chains, deep trenches, and volcanic activity?

A. Mid-ocean ridges
B. Continental shelves
C. Abyssal plains
D. Guyots

QUESTION 224

What is the primary cause of ocean currents?

A. Wind
B. Temperature differences
C. Salinity variations
D. Earth's magnetic field

QUESTION 225

Which marine resource is considered renewable and sustainable when managed properly?

A. Oil deposits
B. Natural gas hydrates
C. Fish
D. Mangrove forests

QUESTION 226

Which of the following is a common erosional feature found along rocky coastlines, formed by the pounding action of waves?

A. Barrier islands
B. Sea stacks
C. Fiords
D. Abyssal plains

QUESTION 227

Which zone of the ocean is known as the "twilight zone" and receives limited sunlight, making it difficult for photosynthesis to occur?

A. Epipelagic zone
B. Mesopelagic zone
C. Bathypelagic zone
D. Abyssopelagic zone

QUESTION 228

Which layer of the atmosphere contains the ozone layer that absorbs harmful ultraviolet (UV) radiation from the sun?

A. Troposphere
B. Stratosphere
C. Mesosphere
D. Thermosphere

QUESTION 229

What is the primary gas responsible for the greenhouse effect in the Earth's atmosphere?

A. Oxygen (O2)
B. Carbon dioxide (CO2)
C. Nitrogen (N2)
D. Methane (CH4)

QUESTION 230

The global wind patterns on Earth are mainly influenced by:

A. Earth's magnetic field
B. Ocean currents
C. The Coriolis effect
D. Human activities

QUESTION 231

The layer of the atmosphere where most weather phenomena occur is the:

A. Troposphere
B. Stratosphere
C. Mesosphere
D. Thermosphere

QUESTION 232

The differential heating of Earth's surface by the sun is responsible for the formation of:

A. Hurricanes
B. Ocean currents
C. Global wind patterns
D. Volcanoes

QUESTION 233

Which gas is a major component of photochemical smog and is produced by vehicular emissions and industrial activities?

A. Oxygen (O2)
B. Carbon dioxide (CO2)
C. Nitrogen dioxide (NO2)
D. Methane (CH4)

QUESTION 234

The release of chlorofluorocarbons (CFCs) into the atmosphere is known to cause:

A. Acid rain
B. Ozone depletion
C. Global warming
D. Landslides

QUESTION 235

The process by which water vapor changes directly into ice crystals without passing through the liquid phase is called:

A. Melting
B. Evaporation
C. Condensation
D. Sublimation

QUESTION 236

Which atmospheric layer contains the ionosphere, where the auroras (Northern and Southern Lights) occur?

A. Troposphere
B. Stratosphere
C. Mesosphere
D. Thermosphere

QUESTION 237

The phenomenon of El Niño is characterized by:

A. Unusually cold ocean temperatures in the Pacific
B. Intensified trade winds in the Atlantic
C. Warmer than average sea surface temperatures in the Pacific
D. Decreased hurricane activity in the Indian Ocean

QUESTION 238

Acid rain is mainly caused by the presence of which two pollutants in the atmosphere?

A. Carbon dioxide (CO2) and sulfur dioxide (SO2)
B. Nitrogen oxides (NOx) and carbon monoxide (CO)
C. Sulfur dioxid (SO2) and nitrogen oxides (NOx)
D. Methane (CH4) and carbon monoxide (CO)

QUESTION 239

The greenhouse effect is beneficial for life on Earth because it:

A. Regulates the ozone layer
B. Prevents hurricanes and tornadoes
C. Traps heat to maintain a habitable temperature
D. Absorbs harmful cosmic radiation

QUESTION 240

Which atmospheric layer is characterized by a decrease in temperature with increasing altitude, known as the "temperature inversion"?

- A. Troposphere
- B. Stratosphere
- C. Mesosphere
- D. Thermosphere

QUESTION 241

What is the primary source of energy that drives the Earth's weather and climate?

- A. Geothermal heat from the Earth's core
- B. Nuclear reactions within the Earth's mantle
- C. Solar radiation from the sun
- D. Cosmic rays from outer space

QUESTION 242

The rapid warming of the Earth's climate in recent decades is primarily attributed to:

- A. Natural fluctuations in solar activity
- B. Changes in the Earth's orbit
- C. Human activities, such as burning fossil fuels
- D. Shifts in the Earth's magnetic field

QUESTION 243

The layer of the atmosphere closest to the Earth's surface is the troposphere. It extends from the surface to approximately 10 kilometers (6 miles) above sea level. The majority of weather phenomena, such as clouds, storms, and precipitation, occur in this layer due to the abundance of water vapor and atmospheric convection. Which layer of the atmosphere is primarily responsible for Earth's weather patterns?

- A. Stratosphere
- B. Troposphere
- C. Mesosphere
- D. Thermosphere

QUESTION 244

The Coriolis effect is a result of Earth's rotation, influencing the direction of moving objects on its surface, including air masses. Due to this effect, air moving toward the poles from the equator is deflected to the east in the Northern Hemisphere and to the west in the Southern Hemisphere. Consequently, this creates the prevailing wind patterns, known as trade winds, westerlies, and polar easterlies. Which factor is primarily responsible for the global wind patterns observed on Earth?

- A. Ocean currents
- B. The Coriolis effect
- C. Earth's magnetic field
- D. Human activities

QUESTION 245

Greenhouse gases in the Earth's atmosphere play a crucial role in maintaining a habitable climate by trapping some of the outgoing infrared radiation, preventing it from escaping into space. However, an excessive increase in the concentration of these gases, such as carbon dioxide (CO_2) and methane (CH_4), has led to an enhanced greenhouse effect, contributing to global warming and climate change. What is the main source of the additional greenhouse gases causing the enhanced greenhouse effect?

    A.   Natural volcanic activity
    B.   Sunspot cycles
    C.   Changes in Earth's orbit
    D.   Human activities, such as burning fossil fuels and deforestation

QUESTION 246

The ozone layer, located in the stratosphere, plays a crucial role in shielding the Earth's surface from harmful ultraviolet (UV) radiation. However, the depletion of the ozone layer has been a major environmental concern. One class of human-made compounds known to be the primary contributors to ozone depletion is chlorofluorocarbons (CFCs). What are CFCs used for and where can they be commonly found?

    A.   Used in agriculture for pest control and found in volcanic emissions
    B.   Used in refrigeration and air conditioning systems and found in aerosol propellants
    C.   Used in water purification and found in naturally occurring hot springs
    D.   Used in manufacturing plastics and found in oceanic algae

QUESTION 247

One of the consequences of climate change is the alteration of global precipitation patterns, which can lead to more intense and frequent droughts in some regions, while others experience heavier rainfall and increased flood risk. This change in precipitation patterns is often linked to the phenomenon known as El Niño-Southern Oscillation (ENSO). What characterizes an El Niño event and how does it impact global weather patterns?

    A.   El Niño refers to abnormally low sea surface temperatures in the Pacific, leading to increased hurricane activity in the Atlantic.
    B.   El Niño refers to a weakening of the trade winds and the warming of sea surface temperatures in the central and eastern Pacific, causing weather disruptions worldwide.
    C.   El Niño refers to the strengthening of the trade winds and the cooling of sea surface temperatures in the Pacific, resulting in more frequent typhoons in the Western Pacific.
    D.   El Niño refers to the increased flow of warm water from the Atlantic Ocean to the Indian Ocean, causing monsoon failures in Southeast Asia.

QUESTION 248

What are the typical weather conditions associated with a high-pressure system?

    A.   Warm temperatures, clear skies, and calm winds
    B.   Cold temperatures, cloudy skies, and heavy rain
    C.   Thunderstorms, strong winds, and low temperatures
    D.   Tornadoes, hailstorms, and high humidity

QUESTION 249

Which type of air mass is responsible for hot and humid weather during the summer in the southeastern United States?

    A.   Polar maritime
    B.   Continental polar
    C.   Maritime tropical
    D.   Continental tropical

QUESTION 250

What type of front occurs when a warm air mass overtakes a cold air mass, lifting the colder air off the ground?

- A. Cold front
- B. Warm front
- C. Stationary front
- D. Occluded front

QUESTION 251

Which of the following clouds are typically associated with thunderstorms?

- A. Cirrus clouds
- B. Cumulonimbus clouds
- C. Stratus clouds
- D. Altocumulus clouds

QUESTION 252

The subtropical jet stream is generally located:

- A. Near the equator
- B. Near the poles
- C. Between the Ferrel and Polar cells
- D. Along the Polar front

QUESTION 253

Which body of water can influence the formation of hurricanes in the Atlantic basin?

- A. Mediterranean Sea
- B. Gulf of Mexico
- C. Arabian Sea
- D. Bay of Bengal

QUESTION 254

What instrument is used to measure atmospheric pressure?

- A. Anemometer
- B. Hygrometer
- C. Barometer
- D. Thermometer

QUESTION 255

Which weather symbol on a weather map represents a cold front?

- A. Red line with triangles
- B. Blue line with half circles
- C. Purple line with alternating triangles and half circles
- D. Black line with crosses

QUESTION 256

When warm air rises and cools, and condensation occurs, what type of clouds are typically formed?

A. Stratus clouds
B. Cirrus clouds
C. Cumulus clouds
D. Nimbostratus clouds

QUESTION 257

Which type of precipitation occurs when raindrops freeze into ice pellets before reaching the ground?

A. Hail
B. Sleet
C. Snow
D. Freezing rain

QUESTION 258

How does the presence of a nearby mountain range affect weather patterns in a region?

A. It causes the region to be warmer and drier.
B. It enhances the intensity of hurricanes.
C. It has no effect on weather patterns.
D. It leads to more frequent tornadoes.

QUESTION 259

Which severe weather phenomenon is characterized by a rotating column of air extending from a thunderstorm to the ground?

A. Hurricane
B. Tornado
C. Blizzard
D. Tsunami

QUESTION 260

Which weather instrument is used to measure wind speed?

A. Anemometer
B. Hygrometer
C. Barometer
D. Wind vane

QUESTION 261

What is the purpose of contour lines on a weather map?

A. To indicate areas of equal temperature
B. To represent areas of equal precipitation
C. To show areas of equal air pressure
D. To display areas of equal wind speed

QUESTION 262

What is the term for a large-scale weather pattern that influences the climate of a region for an extended period, such as El Niño or La Niña?

- A. Monsoon
- B. Tornado
- C. Drought
- D. Climate oscillation

QUESTION 263

Which of the following weather phenomena typically form under the conditions of a low-pressure system?

- A. Thunderstorms occur in low-pressure systems due to the rising warm, moist air that creates unstable atmospheric conditions, leading to the formation of cumulonimbus clouds and potential severe weather, including lightning, heavy rain, and sometimes tornadoes.
- B. Clear skies and calm winds are commonly observed in low-pressure systems because descending cool air suppresses cloud formation and results in stable atmospheric conditions, preventing the development of significant weather events.
- C. Low-pressure systems are associated with cool temperatures, overcast skies, and light precipitation, typically in the form of drizzle or light rain, due to the convergence of different air masses and the absence of strong temperature gradients.
- D. Low-pressure systems often lead to dry and arid conditions, as the descending air warms up and evaporates moisture, resulting in minimal cloud cover and limited chances of precipitation.

QUESTION 264

Which type of air mass is responsible for bringing cold and dry weather to the central regions of a continent during the winter months?

- A. Polar maritime air masses originating over cold oceanic regions transport their moisture over land, leading to cold and dry conditions in the central regions of a continent during winter.
- B. Continental tropical air masses originating from warm desert regions transport their dry and warm air over land, resulting in cold and dry weather in the central regions of a continent during winter.
- C. Maritime polar air masses originating over warm oceanic regions transport their moisture over land, leading to cold and dry conditions in the central regions of a continent during winter.
- D. Arctic air masses originating from the polar regions transport extremely cold and dry air over land, resulting in frigid and dry weather in the central regions of a continent during winter.

QUESTION 265

What type of front is formed when a warm air mass and a cold air mass stall and remain nearly stationary for an extended period?

- A. Cold front
- B. Warm front
- C. Stationary front
- D. Occluded front

QUESTION 266

What type of clouds are typically associated with high-altitude winds and often form in parallel bands or ribbons?

- A. Stratus clouds
- B. Cirrus clouds
- C. Cumulonimbus clouds
- D. Nimbostratus clouds

QUESTION 267

How does the presence of a large body of water, such as an ocean or a large lake, influence the climate of a coastal region?

A. Large bodies of water cause coastal regions to experience more extreme temperature variations, with hotter summers and colder winters, due to the water's slow heating and cooling properties.

B. Coastal regions near large bodies of water tend to have milder and more moderate climates compared to inland areas because water has a higher heat capacity and acts as a temperature regulator, moderating temperature extremes.

C. The presence of large bodies of water leads to a decrease in precipitation in coastal regions, as water evaporates from the surface, causing rain clouds to dissipate before reaching the coast.

D. Coastal regions near large bodies of water experience a significant increase in thunderstorm activity due to the high humidity and temperature differences between the land and water surfaces.

QUESTION 268

What is the primary gas responsible for Earth's greenhouse effect?

A. Oxygen
B. Nitrogen
C. Carbon dioxide
D. Argon

QUESTION 269

Which layer of the atmosphere contains the ozone layer?

A. Troposphere
B. Stratosphere
C. Mesosphere
D. Thermosphere

QUESTION 270

Which weather instrument is used to measure atmospheric pressure?

A. Anemometer
B. Barometer
C. Hygrometer
D. Thermometer

QUESTION 271

The Coriolis effect influences the movement of objects on Earth's surface due to:

A. Earth's rotation
B. Atmospheric pressure
C. Magnetic fields
D. Solar radiation

QUESTION 272

What type of cloud is associated with thunderstorms and heavy rainfall?

A. Cirrus
B. Cumulonimbus
C. Stratus
D. Altostratus

QUESTION 273

The boundary between two air masses with different temperatures and humidity is called a:

A. Front
B. Tropopause
C. Convergence zone
D. Jet strear

QUESTION 274

El Niño and La Niña are climatic phenomena influenced by:

A. Solar flares
B. Earth's tilt
C. Ocean currents
D. Volcanic eruptions

QUESTION 275

The greenhouse effect is a natural process, but human activities have intensified it primarily by:

A. Reducing deforestation
B. Increasing volcanic eruptions
C. Burning fossil fuels
D. Encouraging renewable energy sources

QUESTION 276

What causes the seasons on Earth?

A. Changes in Earth's distance from the Sun
B. Variation in Earth's axial tilt
C. Changes in the Sun's intensity
D. Shifting of the Earth's magnetic poles

QUESTION 277

Which layer of the atmosphere is closest to the Earth's surface?

A. Thermosphere
B. Stratosphere
C. Troposphere
D. Mesosphere

QUESTION 278

Which weather phenomenon is characterized by a rotating column of air extending from a thunderstorm to the ground?

A. Tornado
B. Hurricane
C. Blizzard
D. Tsunami

QUESTION 279

Which climate zone is known for its hot and arid conditions, with little vegetation?

A. Tropical rainforest
B. Desert
C. Tundra
D. Temperate grassland

QUESTION 280

The Earth's atmosphere is primarily composed of:

A. Oxygen and nitrogen
B. Carbon dioxide and oxygen
C. Water vapor and nitrogen
D. Argon and carbon dioxide

QUESTION 281

Which weather instrument is used to measure wind speed?

A. Hygrometer
B. Barometer
C. Anemometer
D. Thermometer

QUESTION 282

The process of water vapor turning directly into ice without passing through the liquid phase is called:

A. Condensation
B. Evaporation
C. Sublimation
D. Precipitation

QUESTION 283

Which of the following is an example of an abiotic characteristic in Earth's major climate regions?

A. Plant species diversity
B. Temperature fluctuations
C. Predator-prey relationships
D. Microbial population density

QUESTION 284

What climate phenomenon is responsible for the periodic warming of sea surface temperatures in the central and eastern equatorial Pacific Ocean?

A. Monsoon
B. El Niño/Southern Oscillation (ENSO)
C. Tsunami
D. Greenhouse effect

QUESTION 285

Which of the following is a significant effect of climate change on ecosystems?

  A. Increase in plant species diversity
  B. Decrease in water scarcity
  C. Disruption of migration patterns
  D. Growth of coral reefs

QUESTION 286

What is a key component of the hydrologic cycle?

  A. Volcanic eruptions
  B. Atmospheric pressure
  C. Plate tectonics
  D. Evaporation

QUESTION 287

Which of the following is considered a biotic characteristic in a terrestrial ecosystem?

  A. Soil type
  B. Annual precipitation
  C. Availability of sunlight
  D. Plant competition

QUESTION 288

What is the primary cause of the monsoon climate?

  A. Earth's axial tilt
  B. Ocean currents
  C. Continental drift
  D. Volcanic activity

QUESTION 289

Which of the following is a current concern related to climate change's impact on human society?

  A. Increased crop yields
  B. Expanded habitable zones
  C. Rising sea levels
  D. Enhanced biodiversity

QUESTION 290

What type of climate region experiences very low temperatures, little precipitation, and strong winds, with the ground often frozen?

  A. Tropical rainforest
  B. Desert
  C. Tundra
  D. Mediterranean

QUESTION 291

Which of the following is an example of a positive feedback mechanism in the context of climate change?

- A. Melting ice leading to increased solar radiation absorption
- B. Deforestation reducing greenhouse gas emissions
- C. Ocean absorption of carbon dioxide decreasing atmospheric CO2 levels
- D. Reduced water vapor in the atmosphere causing cooling

QUESTION 292

Which climate region is characterized by hot and dry conditions, and is often associated with cacti and other succulent plants?

- A. Tropical savanna
- B. Temperate grassland
- C. Mediterranean
- D. Desert

QUESTION 293

How do ocean currents influence regional climate?

- A. By causing earthquakes and volcanic eruptions
- B. By affecting the amount of precipitation in a region
- C. By creating hurricanes and typhoons
- D. By redistributing heat around the globe

QUESTION 294

What is a likely consequence of the disruption of the thermohaline circulation in the Atlantic Ocean?

- A. Reduced precipitation in surrounding regions
- B. Increased frequency of tornadoes
- C. Enhanced fish populations
- D. Altered weather patterns in Europe

QUESTION 295

Which human activity is a significant contributor to the increase in greenhouse gas concentrations in the atmosphere?

- A. Reforestation efforts
- B. Wind energy production
- C. Fossil fuel combustion
- D. Marine conservation

QUESTION 296

In the context of global climate change, what is the "greenhouse effect"?

- A. The cooling of Earth's surface due to volcanic eruptions
- B. The expansion of forested areas on the planet
- C. The trapping of heat in the atmosphere by certain gases
- D. The migration of animals to cooler regions

QUESTION 297

What is a potential consequence of a weakened ozone layer in the stratosphere?

A. Decreased greenhouse gas emissions
B. Increased incidence of skin cancer
C. Enhanced visibility of celestial objects
D. Reduced storm intensity

QUESTION 298

What is the primary gas that makes up Earth's atmosphere?

A. Nitrogen
B. Oxygen
C. Carbon dioxide
D. Argon

QUESTION 299

Which layer of the atmosphere is responsible for protecting life on Earth by absorbing harmful ultraviolet radiation?

A. Troposphere
B. Stratosphere
C. Mesosphere
D. Thermosphere

QUESTION 300

What is the boundary between the troposphere and the stratosphere called?

A. Tropopause
B. Stratosphere transition
C. Mesopause
D. Thermopause

QUESTION 301

What type of weather system is characterized by low pressure at its center and rotating winds?

A. Tornado
B. Hurricane
C. Thunderstorm
D. Blizzard

QUESTION 302

Which weather condition occurs when warm air rises, cools, and condenses to form clouds and precipitation?

A. Fog
B. Hail
C. Front
D. Rain

QUESTION 303

What type of cloud is wispy, high-altitude, and consists of ice crystals?

A. Cumulus
B. Cirrus
C. Stratus
D. Nimbostratus

QUESTION 304

Which gas is primarily responsible for the greenhouse effect, trapping heat in Earth's atmosphere?

A. Carbon dioxide
B. Oxygen
C. Nitrogen
D. Methane

QUESTION 305

What natural phenomenon, caused by changes in Earth's orbit, affects climate by producing variations in solar radiation received on Earth's surface?

A. El Niño
B. Solar flares
C. Milankovitch cycles
D. La Niña

QUESTION 306

Which region near the equator is characterized by calm winds and low atmospheric pressure, known for its typically rainy weather?

A. Polar front
B. Subtropical ridge
C. Intertropical convergence zone (ITCZ)
D. Trade winds

QUESTION 307

Which layer of the atmosphere is closest to the Earth's surface and is where weather phenomena, such as clouds, rain, and thunderstorms, primarily occur?

A. Troposphere: The troposphere is the layer closest to the Earth's surface and extends up to about 10-15 kilometers. It contains most of the atmosphere's water vapor, making it the region where weather events take place.
B. Stratosphere: The stratosphere lies above the troposphere and houses the ozone layer, which absorbs harmful ultraviolet radiation. It does not play a significant role in weather phenomena.
C. Mesosphere: The mesosphere is the third layer of the atmosphere, located above the stratosphere. It is too high for most weather events to occur, and temperatures drop rapidly with altitude.
D. Thermosphere: The thermosphere is the outermost layer of the atmosphere and experiences extremely high temperatures due to absorption of solar radiation. Weather phenomena do not occur here.

QUESTION 308

What are the two most abundant gases in Earth's atmosphere, making up approximately 99% of its composition?

A. Nitrogen and oxygen: Nitrogen accounts for about 78% of the atmosphere, while oxygen makes up about 21% of the composition. Together, they form the majority of Earth's atmosphere.
B. Carbon dioxide and methane: While these gases are essential for the greenhouse effect and regulating climate, they are present in relatively small amounts compared to nitrogen and oxygen.
C. Water vapor and argon: Water vapor is a variable component of the atmosphere and can range from almost 0% to 4% by volume. Argon, an inert gas, makes up about 0.93% of the atmosphere.
D. Helium and neon: These noble gases are trace elements in the atmosphere, present in very small quantities and have negligible effects on atmospheric properties.

QUESTION 309

What is a front in meteorology, and what type of weather conditions are associated with it?

A. A front is a boundary between two air masses with different temperatures and humidity levels. Weather conditions associated with fronts include rain, thunderstorms, and changes in wind direction.
B. A front is a term used to describe extreme weather conditions, such as hurricanes and tornadoes. It is not related to air masses and their interactions.
C. A front is the center of low pressure in a cyclonic system, responsible for creating fair weather conditions with clear skies and gentle winds.
D. A front is the zone of high pressure that leads to sinking air and stable weather conditions with little or no precipitation.

QUESTION 310

What is the greenhouse effect, and how does it impact Earth's climate?

A. The greenhouse effect is a natural process that allows sunlight to pass through Earth's atmosphere and warm its surface. As the Earth radiates heat back into space, greenhouse gases trap some of this heat, preventing excessive cooling and maintaining a stable climate.
B. The greenhouse effect is a human-induced phenomenon caused by excessive deforestation, leading to a reduction in the number of trees that absorb carbon dioxide. This results in an accumulation of greenhouse gases and contributes to global warming.
C. The greenhouse effect is a process where Earth's atmosphere becomes transparent to incoming solar radiation but absorbs and traps heat radiated from the Earth's surface. This leads to rising temperatures and climate change.
D. The greenhouse effect is an atmospheric condition where Earth's temperature is regulated by the presence of greenhouse gases, which absorb and re-emit infrared radiation. It is essential for maintaining a habitable climate on Earth.

QUESTION 311

What is the role of ocean currents in influencing Earth's climate, especially in coastal regions?

A. Ocean currents act as natural air conditioners, absorbing heat from the atmosphere near coastal areas during summers and releasing it back during winters, thereby moderating the temperature.
B. Ocean currents play a minimal role in climate and primarily affect local weather patterns rather than broader climatic conditions.
C. Ocean currents transport massive amounts of water, redistributing heat around the globe, which can significantly influence weather and climate patterns on both regional and global scales.
D. Ocean currents have no impact on Earth's climate, as their influence is limited to water movement and has no connection with atmospheric processes.

QUESTION 312

What is the name of the phenomenon when the Moon passes between the Earth and the Sun, blocking the Sun's light?

A. Solar Eclipse
B. Lunar Eclipse
C. Equinox
D. Aurora Borealis

QUESTION 313

Which planet in our solar system is known as the "Red Planet"?

A. Mercury
B. Venus
C. Mars
D. Jupiter

QUESTION 314

What is the name of the region beyond Neptune where many small icy bodies are located?

A. Kuiper Belt
B. Oort Cloud
C. Asteroid Belt
D. Hubble Zone

QUESTION 315

Which star is located at the center of our solar system?

A. Betelgeuse
B. Polaris
C. Sirius
D. Sun

QUESTION 316

What causes the changing phases of the Moon as observed from Earth?

A. Earth's shadow
B. Solar flares
C. Moon's atmosphere
D. Moon's position relative to the Sun and Earth

QUESTION 317

Which planet is known for its prominent ring system?

A. Mars
B. Saturn
C. Uranus
D. Neptune

QUESTION 318

What is the largest moon in our solar system, with a diameter even larger than the planet Mercury?

A. Europa
B. Titan
C. Io
D. Encolour

QUESTION 319

Which astronomer formulated the three laws of planetary motion, describing the motion of planets around the Sun?

A. Galileo Galilei
B. Johannes Kepler
C. Nicolaus Copernicus
D. Isaac Newton

QUESTION 320

The "Great Red Spot" is a persistent storm observed on which planet?

A. Jupiter
B. Mars
C. Saturn
D. Uranus

QUESTION 321

What is the process by which a star converts hydrogen into helium, releasing an enormous amount of energy in the process?

A. Stellar nucleosynthesis
B. Nuclear fission
C. Nuclear fusion
D. Stellar accretion

QUESTION 322

What term describes the point in a planet's orbit where it is farthest from the Sun?

A. Aphelion
B. Perihelion
C. Equinox
D. Solstice

QUESTION 323

What is the name of the galaxy that contains our solar system?

A. Andromeda Galaxy
B. Milky Way Galaxy
C. Sombrero Galaxy
D. Pinwheel Galaxy

QUESTION 324

What type of star is formed when a massive star undergoes a supernova explosion and collapses under its gravity?

A. White dwarf
B. Red giant
C. Neutron star
D. Black hole

QUESTION 325

What is the name of the imaginary line around which Earth rotates, resulting in day and night?

A. Equator
B. Prime Meridian
C. Tropic of Cancer
D. Axis

QUESTION 326

Which of Jupiter's moons is known for its subsurface ocean, making it a potential candidate for extraterrestrial life?

A. Ganymede
B. Callisto
C. Europa
D. Io

QUESTION 327

What is the primary source of energy production in the Sun?

A. Nuclear fusion of hydrogen into helium in the core
B. Nuclear fission of heavy elements in the outer layers
C. Combustion of gases in the Sun's atmosphere
D. Nuclear fusion of helium into heavier elements in the corona

QUESTION 328

Which of the following is a characteristic of a main-sequence star?

A. It undergoes rapid changes in size due to frequent supernovas.
B. It is the final stage in a star's life cycle.
C. It fuses helium into heavier elements during its evolution.
D. It maintains a stable balance between gravity and nuclear fusion.

QUESTION 329

What process is responsible for the nucleosynthesis of elements heavier than iron in a massive star's core?

A. Solar flares
B. Stellar winds
C. Supernova explosion
D. Black hole formation

QUESTION 330

Which type of galaxy is characterized by a flattened, disk-like shape, and typically contains a central bulge and spiral arms?

A. Elliptical galaxy
B. Irregular galaxy
C. Spiral galaxy
D. Lenticular galaxy

QUESTION 331

What is the prevailing scientific theory regarding the origin of the universe?

- A. Steady State Theory
- B. Big Bang Theory
- C. Geocentric Theory
- D. Ptolemaic Theory

QUESTION 332

Which of the following is a characteristic of a black hole?

- A. Emits a large amount of visible light
- B. Its gravity pulls matter inward but never releases it
- C. Forms from the fusion of two neutron stars
- D. Exists in the center of every galaxy

QUESTION 333

Which type of technology is used to detect and study distant objects in space by analyzing the different wavelengths of electromagnetic radiation they emit?

- A. Spectroscopy
- B. Radiocarbon dating
- C. Sonar technology
- D. Magnetic resonance imaging (MRI)

QUESTION 334

What is the name given to planets that exist outside our solar system and orbit a star?

- A. Exoplanets
- B. Interplanetary planets
- C. Celestial planets
- D. Intra-solar planets

QUESTION 335

What process powers the Sun and other stars, converting hydrogen into helium and releasing energy?

- A. Gravitational collapse
- B. Nuclear fission
- C. Nuclear fusion
- D. Stellar accretion

QUESTION 336

Which of the following is not a characteristic of dark matter?

- A. It does not emit, absorb, or reflect light.
- B. It interacts with gravity and can be detected through its gravitational effects.
- C. It is composed of ordinary matter, such as atoms and molecules.
- D. It makes up a significant portion of the total mass in the universe.

QUESTION 337

What is the final explosive stage in the life cycle of a massive star, resulting in a brilliant burst of light before it collapses into either a neutron star or a black hole?

    A.   Stellar wind
    B.   Nova
    C.   Supernov
    D.   Solar flare

QUESTION 338

Which of the following is a high-energy astronomical object characterized by an incredibly bright core powered by the accretion of material onto a supermassive black hole?

    A.   Quasar
    B.   Pulsar
    C.   Nebula
    D.   Cepheid variable star

QUESTION 339

Which space telescope, launched by NASA in 1990, has provided astronomers with valuable images and data about the universe across various wavelengths?

    A.   Kepler Space Telescope
    B.   Hubble Space Telescope
    C.   Spitzer Space Telescope
    D.   Chandra X-ray Observatory

QUESTION 340

What is the process by which a massive star becomes a black hole at the end of its life cycle?

    A.   Stellar explosion
    B.   Gravitational collapse
    C.   Fusion ignition
    D.   Solar convection

Answer:

QUESTION 341

What type of galaxy has a smooth, nearly spherical shape and is characterized by the absence of spiral arms?

    A.   Spiral galaxy
    B.   Lenticular galaxy
    C.   Elliptical galaxy
    D.   Irregular galaxy

QUESTION 342

Suppose astronomers observe a star that exhibits a periodic increase and decrease in brightness. What could be a potential explanation for this behavior?

    A.   The star is undergoing nuclear fusion in its core.
    B.   The star is experiencing stellar winds causing fluctuations in brightness.
    C.   The star is part of a binary star system, and its brightness changes as it orbits its companion.
    D.   The star is a pulsar, emitting regular bursts of radiation.

QUESTION 343

Imagine a distant galaxy appears to have a redshift in its spectral lines. What does this redshift imply about the motion of the galaxy?

    A. The galaxy is moving toward Earth.
    B. The galaxy is stationary, not moving relative to Earth.
    C. The galaxy is moving away from Earth.
    D. The galaxy's motion cannot be determined based on the redshift.

QUESTION 344

Suppose an astronomer discovers a new celestial object with an extremely high gravitational pull that even light cannot escape from it. What could this object likely be?

    A. A white dwarf
    B. A black hole
    C. A pulsar
    D. A red giant

QUESTION 345

A star's luminosity is a measure of:

    A. Its apparent brightness as observed from Earth.
    B. The amount of visible light it emits.
    C. Its size and mass relative to other stars.
    D. The total amount of energy it radiates per unit time.

QUESTION 346

A telescope observes a distant galaxy and finds that it has a bluer color compared to nearby galaxies. What could be a possible explanation for this observation?

    A. The galaxy is very old, and its stars have exhausted their fuel.
    B. The galaxy is experiencing a period of intense star formation.
    C. The galaxy is moving away from us at a high speed.
    D. The telescope's lenses are defective, causing a blue shift in colors.

QUESTION 347

Which planet in our solar system is known as the "Red Planet" due to its reddish appearance?

    A. Venus
    B. Mars
    C. Jupiter
    D. Uranus

QUESTION 348

What is the largest planet in our solar system?

    A. Saturn
    B. Neptune
    C. Jupiter
    D. Mercury

QUESTION 349

Which celestial object is made up of dust, rock, and ice, and typically develops a bright glowing tail when close to the Sun?

A. Meteor
B. Comet
C. Asteroid
D. Moon

QUESTION 350

Which planet in our solar system has the most prominent ring system?

A. Uranus
B. Saturn
C. Neptune
D. Jupiter

QUESTION 351

What is the largest moon in our solar system and orbits the planet Jupiter?

A. Titan
B. Ganymede
C. Europa
D. Io

QUESTION 352

Which planet has the most rapid rotation, completing one rotation in about 10 hours?

A. Venus
B. Mars
C. Jupiter
D. Saturn

QUESTION 353

What is the name of the region between Mars and Jupiter that is populated with irregularly shaped objects, most of which are found in the "asteroid belt"?

A. Kuiper Belt
B. Oort Cloud
C. Taurid Complex
D. Asteroid Belt

QUESTION 354

Which mathematical law describes the relationship between a planet's orbital period and its distance from the Sun?

A. Kepler's First Law
B. Kepler's Second Law
C. Kepler's Third Law
D. Newton's Law of Universal Gravitation

QUESTION 355

What is the name of the largest volcano in our solar system, located on the planet Mars?

    A.   Mount Vesuvius
    B.   Mauna Loa
    C.   Olympus Mons
    D.   Mount St. Helens

QUESTION 356

Which planet in our solar system experiences the phenomenon known as the "Great Red Spot"?

    A.   Jupiter
    B.   Saturn
    C.   Neptune
    D.   Uranus

QUESTION 357

What is the name given to the process by which a comet's icy nucleus vaporizes and forms a glowing atmosphere when it gets close to the Sun?

    A.   Sublimation
    B.   Fusion
    C.   Ionization
    D.   Outgassing

QUESTION 358

Which planet has the most extreme temperatures in our solar system, with scorching daytime temperatures and freezing nighttime temperatures?

    A.   Venus
    B.   Mars
    C.   Mercury
    D.   Uranus

QUESTION 359

What is the name of the spacecraft that successfully landed on Saturn's largest moon, Titan, in 2005, and discovered lakes and seas of liquid hydrocarbons on its surface?

    A.   Voyager 1
    B.   Cassini-Huygens
    C.   New Horizons
    D.   Juno

QUESTION 360

Which mathematical law describes the relationship between a planet's average distance from the Sun and the time it takes to complete one orbit around the Sun?

    A.   Kepler's First Law
    B.   Kepler's Second Law
    C.   Kepler's Third Law
    D.   Newton's Law of Universal Gravitation

QUESTION 361

What is the name of the most massive asteroid in the asteroid belt between Mars and Jupiter?

- A. Ceres
- B. Vesta
- C. Eros
- D. Pallas

QUESTION 362

Suppose a student asks why some planets in our solar system have rings while others do not. Which explanation would be most appropriate for the teacher to provide?

- A. "Planets with rings are closer to the Sun and receive more sunlight, causing the formation of rings."
- B. "The presence of rings is purely random and does not depend on any specific factors."
- C. "Planets with rings have stronger gravitational forces, which trap debris and create ring systems."
- D. "Rings are remnants of ancient moons that disintegrated due to collisions with asteroids."

QUESTION 363

A student wonders why Earth's moon looks much larger when it is near the horizon compared to when it is high in the sky. Which explanation would be most appropriate for the teacher to give?

- A. "This is an optical illusion caused by Earth's atmosphere magnifying the moon near the horizon."
- B. "The moon's distance from Earth is closer when it is near the horizon, making it appear larger." (Correct)
- C. "The moon's gravitational pull is stronger when it is near the horizon, causing it to expand in size."
- D. "The moon's reflective properties change as it moves across the sky, creating the illusion of size difference."

QUESTION 364

A student asks why the surface temperature of Venus is much hotter than that of Mercury, even though Mercury is closer to the Sun. Which explanation would be most appropriate for the teacher to provide?

- A. "Venus has a thicker atmosphere that traps more heat, making its surface hotter than Mercury." (Correct)
- B. "Mercury's surface is covered in highly reflective materials, which bounce away most of the Sun's heat."
- C. "The Sun's energy is stronger at Venus, leading to higher temperatures despite its farther distance."
- D. "Mercury's proximity to the Sun causes its surface to be too hot for any heat to accumulate."

QUESTION 365

A student wonders why some comets have highly elongated, elliptical orbits that bring them close to the Sun and then far away into the outer solar system. Which explanation would be most appropriate for the teacher to give?

- A. "Comets' orbits are influenced by the gravitational pull of the outer planets, causing them to elongate."
- B. "The Sun's heat causes comets to change their orbits, resulting in elongated paths."
- C. "Comets originate from different regions of the solar system, which affects their orbital shapes."
- D. "Comets have elliptical orbits because they are affected by the Sun's gravity more at some points in their path."

QUESTION 366

A student asks why the gas giants (Jupiter, Saturn, Uranus, Neptune) have many more moons compared to the terrestrial planets (Mercury, Venus, Earth, Mars), which have only a few moons each. Which explanation would be most appropriate for the teacher to provide?

- A. "The gas giants captured many moons from other solar systems, whereas the terrestrial planets did not."
- B. "Moons are formed from the debris surrounding the gas giants during their initial formation."
- C. "The ga giants' strong gravitational forces allowed them to accumulate more moons over time."
- D. "The terrestrial planets have more moons, but they are too small to observe from Earth."

QUESTION 367

What is the leading scientific theory about the origin of the Moon?

    A.   Collision Theory: The Moon formed when a Mars-sized object collided with Earth, ejecting debris that eventually accreted to form the Moon.

    B.   Capture Theory: The Moon was formed elsewhere in the solar system and was captured by Earth's gravity.

    C.   Fission Theory: The Moon was once part of Earth and separated due to rapid spinning.

    D.   Condensation Theory: The Moon formed from the same nebula of gas and dust as Earth.

QUESTION 368

What is the primary reason for the Moon's tidal effects on Earth?

    A.   The Moon's gravitational pull is stronger on the side of Earth facing away from it.

    B.   The Moon's orbital speed affects Earth's rotational speed.

    C.   The Moon's mass causes a distortion in space-time around Earth.

    D.   The Moon's gravitational force causes a bulge of water on the side facing it.

QUESTION 369

What is the apparent path of the Sun across the sky over the course of a day?

    A.   Equator

    B.   Meridian

    C.   Celestial Equator

    D.   Ecliptic

QUESTION 370

How does the tilt of Earth's axis affect the seasons?

    A.   The tilt causes variations in the distance between Earth and the Sun throughout the year.

    B.   The tilt determines the amount of daylight hours in a day.

    C.   The tilt causes variations in the angle at which sunlight strikes the Earth's surface.

    D.   The tilt affects the speed of Earth's rotation.

QUESTION 371

What causes the phenomenon known as "precession" of Earth's axis?

    A.   Gravitational pull from the Moon and the Sun on Earth's equatorial bulge.

    B.   Changes in the Sun's magnetic field.

    C.   Variations in the Earth's rotation speed.

    D.   Impact of large celestial bodies on Earth's poles.

QUESTION 372

What is the apparent motion of planets relative to the stars from Earth's perspective?

    A.   Retrograde motion

    B.   Direct motion

    C.   Tidal motion

    D.   Equatorial motion

QUESTION 373

What is the term for the point in the Moon's orbit where it is closest to Earth?

A. Apogee
B. Perihelion
C. Aphelion
D. Perigee

QUESTION 374

Which layer of the Earth's atmosphere is responsible for the absorption of most of the ultraviolet (UV) radiation from the Sun?

A. Mesosphere
B. Stratosphere
C. Troposphere
D. Thermosphere

QUESTION 375

Which celestial event occurs when the Moon passes directly between the Sun and Earth, casting a shadow on Earth's surface?

A. Solar eclipse
B. Lunar eclipse
C. Equinox
D. Solstice

QUESTION 376

What causes the phenomenon of "seasonal lag" where the warmest temperatures are often observed after the summer solstice and the coldest temperatures after the winter solstice?

A. Ocean currents
B. Atmospheric pressure changes
C. Heat absorption and release by the Earth's surface
D. Changes in Earth's distance from the Sun

QUESTION 377

Why do some regions near the equator experience relatively consistent temperatures throughout the year, while regions near the poles experience significant seasonal temperature variations?

A. Equatorial regions receive more direct sunlight, leading to consistent temperatures.
B. Polar regions have higher albedo, reflecting more sunlight and causing temperature fluctuations.
C. Equatorial regions experience stronger ocean currents, moderating temperatures.
D. Polar regions are affected by the tilt of Earth's axis, leading to seasonal changes.

QUESTION 378

During a total solar eclipse, why does the Moon appear to completely cover the Sun, despite the Sun's much larger size?

A. The Moon's distance from Earth varies, allowing it to precisely cover the Sun during an eclipse.
B. The Moon's size changes due to gravitational forces during the eclipse.
C. The Moon's orbit aligns perfectly with the Sun and Earth, leading to complete coverage.
D. The Sun's brightness decreases during the eclipse, making it appear smaller.

QUESTION 379

How do ocean tides differ from waves, and what causes their regular rise and fall?

    A. Tides are caused by the gravitational pull of celestial bodies, while waves are caused by wind.
    B. Tides are a result of Earth's rotation, while waves are influenced by the Moon's gravity.
    C. Tides are driven by the Moon's gravitational pull, while waves are generated by earthquakes.
    D. Tides are influenced by the Sun's gravity, while waves are created by ocean currents.

QUESTION 380

Why do some planets, like Mars, have more noticeable retrograde motion compared to other planets like Jupiter or Saturn?

    A. Mars has a larger distance from the Sun, leading to greater retrograde motion.
    B. Mars has a thicker atmosphere, causing its motion to appear more pronounced.
    C. Retrograde motion is an optical illusion observed only from Earth, not from other planets.
    D. Retrograde motion is more common in smaller, rocky planets like Mars.

QUESTION 381

How does the greenhouse effect contribute to the Earth's overall climate and temperature regulation?

    A. The greenhouse effect traps solar energy, causing Earth's temperature to rise.
    B. Greenhouse gases absorb infrared radiation, preventing the escape of heat from Earth.
    C. The greenhouse effect occurs due to deforestation, leading to an increase in CO_2 levels.
    D. Greenhouse gases reflect sunlight, maintaining a stable temperature on Earth.

QUESTION 382

Which of the following is a characteristic of stars?

    A. They emit only visible light.
    B. They are all located in the Milky Way galaxy.
    C. Their color is solely determined by their temperature.
    D. They have constant luminosity throughout their lifetimes.

QUESTION 383

Which of the following statements about galaxies is true?

    A. All galaxies have a similar shape and size.
    B. The Milky Way is the largest galaxy in the universe.
    C. Galaxies contain mostly stars and no interstellar matter.
    D. Galaxies can be grouped into different types based on their shapes.

QUESTION 384

What is the main component of the solar system?

    A. The Moon
    B. Jupiter
    C. The Sun
    D. Saturn

QUESTION 385

Which of the following is NOT a characteristic of the Sun?

- A. It is composed mostly of hydrogen and helium.
- B. It undergoes nuclear fusion in its core.
- C. It has a solid surface that we can land on.
- D. It emits solar flares and solar winds.

QUESTION 386

What celestial event occurs when the Moon passes between the Earth and the Sun?

- A. Lunar eclipse
- B. Solar eclipse
- C. Harvest moon
- D. Blue moon

QUESTION 387

Which of the following statements is true regarding the apparent motion of stars and planets?

- A. Stars and planets move in a straight line across the sky.
- B. The apparent motion of stars is caused by their physical motion relative to Earth.
- C. All stars and planets appear to move at the same speed across the sky.
- D. The apparent motion of stars and planets is an illusion; they are actually stationary.

QUESTION 388

What is the term for the path that the Sun appears to take across the celestial sphere over the course of a year?

- A. Orbit
- B. Rotation
- C. Equator
- D. Ecliptic

QUESTION 389

Which of the following best describes a planet's "opposition" in the night sky?

- A. The planet is at its closest distance to Earth.
- B. The planet is directly above one of the poles.
- C. The planet is in the same direction as the Sun and cannot be observed.
- D. The planet is directly opposite the Sun, rising as the Sun sets.

QUESTION 390

Which of the following factors primarily determines a star's life cycle and ultimate fate?

- A. Its distance from Earth
- B. Its color and apparent brightness
- C. Its mass
- D. Its distance from other stars

QUESTION 391

What is the process responsible for the Sun's energy production?

    A.   Nuclear fission
    B.   Nuclear fusion
    C.   Radioactive decay
    D.   Chemical combustion

QUESTION 392

Which type of star is known for its relatively small size, high density, and incredibly strong gravitational pull?

    A.   White dwarf
    B.   Red giant
    C.   Neutron star
    D.   Blue supergiant

QUESTION 393

What is the term for a sudden and significant increase in a star's brightness?

    A.   Nova
    B.   Comet
    C.   Solar flare
    D.   Meteor

QUESTION 394

Which of the following objects is composed mainly of dust, ice, and gas and forms a bright coma and tail when it approaches the Sun?

    A.   Asteroid
    B.   Meteoroid
    C.   Comet
    D.   Meteor

QUESTION 395

What is the name of the galaxy that contains our solar system?

    A.   Andromeda Galaxy
    B.   Milky Way Galaxy
    C.   Whirlpool Galaxy
    D.   Sombrero Galaxy

QUESTION 396

What is the term for the study of the universe, its origins, and its evolution?

    A.   Astrophysics
    B.   Cosmology
    C.   Geology
    D.   Meteorology

QUESTION 397

Suppose an astronomer observes a star that appears to be red in color. What can the astronomer infer about the star based on its color alone?

A. The star is relatively cooler compared to other stars.
B. The star is relatively hotter compared to other stars.
C. The star is moving away from Earth.
D. The star is moving closer to Earth.

QUESTION 398

During a lunar eclipse, the Earth comes between the Sun and the Moon, causing the Moon to appear darkened. Why is the Moon not completely invisible during a lunar eclipse?

A. The Moon's surface reflects sunlight, making it visible.
B. The Earth's atmosphere scatters sunlight onto the Moon.
C. The Sun's corona emits light, illuminating the Moon.
D. The Moon emits its own light, making it visible.

QUESTION 399

Suppose a distant planet is located in our solar system beyond Neptune. How would its orbital period compare to that of Earth?

A. The planet's orbital period would be shorter than Earth's.
B. The planet's orbital period would be longer than Earth's.
C. The planet's orbital period would be the same as Earth's.
D. The planet would not have an orbital period.

QUESTION 400

Two stars have the same absolute magnitude, but one appears dimmer in the night sky than the other when observed from Earth. What can be inferred about these stars?

A. The dimmer star is farther away from Earth.
B. The dimmer star has a lower luminosity than the brighter star.
C. The dimmer star is undergoing a supernova explosion.
D. The dimmer star has a higher surface temperature than the brighter star.

QUESTION 401

Imagine two planets with the same mass and size, but one is composed mainly of iron and the other of water ice. How would their densities compare?

A. The planet made of iron would have a higher density.
B. The planet made of water ice would have a higher density.
C. Both planets would have the same density.
D. The density cannot be determined with the given information.

QUESTION 402

Which of the following is a key principle for designing and conducting an inquiry-based scientific investigation?

A. Memorizing existing evidence
B. Relying solely on theoretical knowledge
C. Proposing definitive answers and explanations
D. Reviewing current knowledge in light of existing evidence

QUESTION 403

As an Earth science teacher, how can you model scientific attitudes such as curiosity and openness to new ideas?

    A. Discouraging questions from students
    B. Avoiding discussions about current scientific discoveries
    C. Encouraging students to explore and ask questions
    D. Limiting class discussions to pre-established topics

QUESTION 404

What is a vital aspect of designing the learning environment to support student inquiries and scientific investigations?

    A. Restricting access to laboratory resources
    B. Limiting participation to select students
    C. Providing ample time, space, and resources for all students to participate
    D. Relying solely on whole-class settings

QUESTION 405

What is a primary role of the Earth science teacher in assisting students with generating and refining scientific questions and hypotheses?

    A. Providing pre-determined answers and hypotheses
    B. Avoiding discussions about the scientific method
    C. Guiding students in generating, refining, and testing questions and hypotheses
    D. Discouraging students from asking too many questions

QUESTION 406

Why is active learning and inquiry-based methods essential in science instruction?

    A. They simplify complex scientific concepts for students
    B. They reduce student engagement and motivation
    C. They promote deeper understanding and critical thinking
    D. They limit the scope of scientific investigations

QUESTION 407

In an earth science class, the teacher is guiding students through an inquiry-based investigation on the water cycle. Students are tasked with developing explanations for why rainfall patterns differ in various regions of the world. They have collected data on temperature, humidity, and wind patterns in different regions.

Which action by the teacher best exemplifies guiding students to identify, refine, and focus scientific ideas and questions in this inquiry-based investigation?

    A. Providing students with pre-made explanations for the rainfall patterns and having them select the most likely one.
    B. Giving students access to a variety of research papers related to the water cycle and asking them to replicate the experiments described.
    C. Facilitating a class discussion where students share their data, observations, and initial explanations for the rainfall patterns.
    D. Asking students to complete a quiz on the water cycle before starting the investigation.

QUESTION 408

An earth science teacher is planning a sequence of learning activities for a unit on plate tectonics. The teacher wants to uncover common misconceptions and challenge students to expand their understanding of the topic. Which learning activity sequence aligns best with the teacher's objectives?

    A.   Starting with a lecture on plate tectonics, followed by a hands-on activity where students model plate movements using clay, and concluding with a quiz on plate boundaries.

    B.   Assigning students to read a chapter on plate tectonics in their textbooks, then showing a documentary on earthquakes and volcanic eruptions, and ending with a group discussion about plate tectonics.

    C.   Beginning with a classroom debate on the theory of plate tectonics, followed by individual research projects on famous earthquakes, and culminating in a student-led presentation about their findings.

    D.   Providing students with a list of key terms related to plate tectonics and asking them to create flashcards for each term, followed by a lab activity where they study rock samples to identify evidence of past tectonic events.

QUESTION 409

During a science fair, an earth science teacher observes a group of students conducting an investigation on soil erosion. The students are making systematic observations and measurements of different soil types exposed to varying water flow rates.

What could the teacher do to further support the students' inquiry-based investigation?

    A.   Provide the students with a detailed procedure to follow, step-by-step, to ensure accurate and consistent measurements.

    B.   Suggest that the students add different soil additives to their experiment to see how they affect erosion rates.

    C.   Ask the students to conduct their investigation using a single soil sample to reduce variability.

    D.   Encourage the students to record any unexpected or unusual observations and discuss them as a group.

QUESTION 410

An earth science teacher is working with a class of middle school students on understanding the concept of renewable and non-renewable resources. The teacher wants to move the students from concrete to more abstract understanding.

Which approach would best help the teacher achieve this goal?

    A.   Showing a video that explains the differences between renewable and non-renewable resources and providing a handout with examples of each type.

    B.   Assigning students a project where they have to create a collage using pictures of various renewable and non-renewable resources.

    C.   Facilitating a class discussion about the impact of human activities on natural resources and how sustainable practices can help preserve them.

    D.   Organizing a field trip to a local renewable energy facility, followed by a debate on the advantages and disadvantages of renewable energy sources.

QUESTION 411

Ms. Johnson is teaching her Earth Science class about plate tectonics. She wants her students to develop scientific questions related to plate movements. What strategy can she use to assist her students in this process?

    A.   Provide the students with a list of pre-determined scientific questions related to plate tectonics.

    B.   Encourage students to brainstorm individually and write down any questions that come to mind.

    C.   Assign specific questions to each student, ensuring they cover all aspects of plate movements.

    D.   Instruct students to memorize existing scientific questions from textbooks.

QUESTION 412

Mr. Smith wants his Earth Science students to analyze different explanations for a recent volcanic eruption. What should he do to help them achieve this goal?

    A.   Provide a single explanation and ask students to memorize it.
    B.   Assign different explanations to each group and ask them to present their findings.
    C.   Present multiple explanations and encourage students to discuss and compare them.
    D.   Request students to skip the analysis and focus only on potential sources of error.

QUESTION 413

Ms. Lee is conducting an inquiry-based scientific investigation with her Earth Science class to study soil erosion. How can she guide her students in making systematic observations and measurements?

    A.   Provide students with a written description of soil erosion and its effects.
    B.   Take the students on a field trip to a location with soil erosion and let them explore freely.
    C.   Instruct students to make random observations and measurements without any guidance.
    D.   Provide clear instructions and tools for specific observations and measurements related to soil erosion.

QUESTION 414

Mr. Anderson is planning a sequence of learning activities for his Earth Science class to teach the concept of climate change. What should be his approach to uncover common misconceptions among students?

    A.   Avoid discussing misconceptions to prevent confusion among students.
    B.   Present students with a set of pre-determined misconceptions to debunk.
    C.   Encourage students to share their ideas about climate change with the class.
    D.   Provide a single perspective on climate change and discourage other viewpoints.

QUESTION 415

Ms. Johnson is an Earth Science teacher who has introduced a new inquiry-based investigation project to her students. She wants to assess their understanding and performance. Which assessment method should she use to evaluate her students' participation and comprehension of the project?

    A.   Multiple-choice test
    B.   Standardized test
    C.   Rubrics
    D.   True/False questions

QUESTION 416

Mr. Smith is an Earth Science teacher who has been teaching a unit on plate tectonics. He wants to design an assessment that aligns with the learning objectives of the unit. What type of assessment should he create?

    A.   Short-answer test
    B.   Performance assessment
    C.   True/False questions
    D.   Self-assessment

QUESTION 417

Ms. Lee is an Earth Science teacher who wants to continually monitor her students' understanding of key concepts throughout the semester. Which assessment method should she employ to achieve this goal?

    A.   Summative assessment
    B.   Self-assessment
    C.   Formal assessment
    D.   Formative assessment

QUESTION 418

Mr. Davis, an Earth Science teacher, is planning an assessment for his class. He wants to make sure the assessment serves its intended purpose effectively. What is the importance of limiting the use of an assessment to its intended purpose?

    A. To make the assessment more challenging
    B. To increase students' motivation to study
    C. To ensure fairness and validity of the assessment
    D. To align the assessment with state standards

QUESTION 419

Which of the following is an important consideration for Earth Science teachers to ensure inclusive and respectful learning environments?

    A. Providing inquiry-based investigations relevant to students' daily lives.
    B. Utilizing a single instructional strategy to cater to all students.
    C. Ignoring students' prior knowledge and experiences while planning activities.
    D. Excluding English-language learners from science curricula and activities.

QUESTION 420

What research-based factors should Earth Science teachers consider while understanding how students develop scientific understanding?

    A. Students' prior knowledge, experience, and attitudes.
    B. Students' height and weight.
    C. Students' favorite sports and hobbies.
    D. Students' preferred learning style.

QUESTION 421

How can Earth Science teachers promote student self-motivation and engagement in their own learning?

    A. Designing instructional materials using situations from students' daily lives.
    B. Assigning tasks that are unrelated to students' interests and experiences.
    C. Avoiding any collaboration among students.
    D. Using only traditional lecture-style instruction.

QUESTION 422

What is a key approach Earth Science teacher can use to ensure all students comprehend content-related texts?

    A. Using a variety of instructional strategies.
    B. Relying solely on a single instructional approach.
    C. Providing only written texts and no other technologies.
    D. Ignoring the diverse needs of students.

QUESTION 423

A new Earth Science teacher wants to assess students' prior knowledge and misconceptions about science at the beginning of the school year. Which strategy would be most effective for the teacher to achieve this goal?

    A. Assign a lengthy written test covering all Earth Science topics.
    B. Conduct an interactive group discussion on various science concepts.
    C. Ask students to create a presentation on a scientific topic of their choice.
    D. Assign a project that requires students to conduct experiments related to Earth Science concepts.

QUESTION 424

An Earth Science teacher wants to evaluate the reliability and validity of an assessment instrument. Which action should the teacher take to assess these characteristics properly?

- A. Administer the assessment to a small group of students instead of the whole class.
- B. Compare the results of the assessment with students' prior test scores in other subjects.
- C. Ask students for their opinions on the difficulty level of the assessment.
- D. Analyze the consistency of scores obtained from multiple administrations of the assessment.

QUESTION 425

An Earth Science teacher wants to engage students in meaningful self-assessment. What strategy would be most effective for achieving this goal?

- A. Providing detailed feedback only during parent-teacher conferences.
- B. Allowing students to grade their own assessments without guidance.
- C. Sharing evaluation criteria with students before the assessment.
- D. Conducting assessments without informing students in advance.

QUESTION 426

An Earth Science teacher believes in providing equal opportunities for all students to demonstrate their achievements. Which assessment method aligns best with this belief?

- A. A multiple-choice test with a strict time limit.
- B. A take-home assignment with flexible deadlines.
- C. A group project with individual grading.
- D. A written test without any opportunity for revisions.

Chapter 2 – Answers and Explanations

QUESTION 1

Answer: B

Explanation: A scientific investigation involves a systematic and organized approach to explore and solve a specific problem or question. It includes making observations, collecting relevant data, analyzing the data, and drawing conclusions based on evidence.

QUESTION 2

Answer: C

Explanation: Organizing and analyzing scientific data allow researchers to identify patterns, trends, and relationships that can provide valuable insights and support or refute hypotheses. It helps draw meaningful conclusions and make informed decisions based on evidence.

QUESTION 3

Answer: C

Explanation: Presenting scientific data in a clear and unbiased manner is crucial to facilitate effective communication and peer review. It allows other scientists to assess the validity of the research and replicate the study if needed, ensuring the reliability of the results.

QUESTION 4

Answer: C

Explanation: Hypotheses are proposed explanations or predictions for a specific phenomenon or problem that are based on existing knowledge or observations. They guide the research and help in designing the investigation to test their validity.

QUESTION 5

Answer: C

Explanation: The evidential basis of scientific claims is rooted in the logical reasoning and empirical evidence derived from systematic observations, data collection, and analysis. It involves presenting evidence to support or refute a hypothesis based on objective, verifiable data.

QUESTION 6

Answer: C

Explanation: Collecting data from multiple sources, when applicable, can enhance the reliability and validity of scientific findings. It allows researchers to consider various perspectives and ensure the results are not limited to one specific source.

QUESTION 7

Answer: B

Explanation: The control group in a scientific investigation is used as a baseline for comparison with the experimental group that receives the treatment or intervention. It helps researchers assess the effectiveness of the treatment by providing a standard of comparison for evaluating the results.

QUESTION 8

Answer: C

Explanation: The scientific method is a systematic approach to scientific investigations that involves formulating a hypothesis, conducting experiments or data collection, analyzing the data, and drawing conclusions based on empirical evidence. Drawing conclusions solely based on intuition lacks the rigor and objectivity required in the scientific process.

QUESTION 9

Answer: C

Explanation: Considering potential sources of bias is essential in scientific data interpretation to acknowledge and control for any factors that may influence the results. This helps maintain the integrity of the research and ensures that conclusions are based on valid and reliable data.

QUESTION 10

Answer: C

Explanation: Peer review is a critical step in the scientific process where independent experts in the field assess and evaluate a research paper's quality, validity, and methodology. It helps identify any errors, weaknesses, or biases in the research and ensures that the conclusions are based on robust evidence before publication.

QUESTION 11

Answer: A

Explanation: Plate tectonics is driven by the movement of molten rock in the Earth's mantle, which creates convection currents. These currents cause the lithospheric plates to move, leading to various geological phenomena such as earthquakes, volcanic activity, and the formation of mountain ranges.

QUESTION 12

Answer: D

Explanation: Oxbow lakes are formed when a meandering river changes its course over time, and a new straighter channel is created, leaving behind a U-shaped lake that is separated from the main river flow.

QUESTION 13

Answer: C

Explanation: Igneous rocks are formed from the cooling and solidification of magma (below the Earth's surface) or lava (on the Earth's surface). Examples of igneous rocks include basalt, granite, and pumice.

QUESTION 14

Answer: D

Explanation: Coastal erosion is primarily caused by the relentless action of waves and currents along the shoreline. Over time, these forces wear away the coastal landforms, leading to the gradual retreat of the coastline.

QUESTION 15

Answer: A

Explanation: Calving is the process by which large blocks of ice break off from the front of a glacier and fall into the ocean, forming icebergs. This is a common occurrence in areas where glaciers flow into the sea, such as tidewater glaciers.

QUESTION 16

Answer: D

Explanation: Safety goggles, lab coats, and gloves are essential PPE in science investigations to protect against chemical spills, splashes, and other potential hazards. Students and teachers must always wear these safety measures to ensure their safety during experiments.

QUESTION 17

Answer: C

Explanation: Mercury is a hazardous material commonly found in thermometers and other laboratory equipment. It is toxic and can be absorbed through the skin or inhaled as a vapor. Extreme care should be taken when handling mercury, and proper disposal procedures must be followed.

QUESTION 18

Answer: C

Explanation: Scientific notation is a standard way of expressing large or small numbers in a compact form, making it easier to handle and compare data. It is commonly used in Earth and space science to represent numbers such as distances, masses, and concentrations.

QUESTION 19

Answer: C

Explanation: Hazardous waste should never be disposed of in sinks, regular trash bins, or mixed together. Instead, it must be collected and stored in properly labeled, leak-proof containers until it can be disposed of safely following local regulations.

QUESTION 20

Answer: C

Explanation: Liters (L) are the appropriate unit for measuring the volume of liquids in Earth and space science investigations. It is a standard unit for volume and is commonly used in experiments involving liquids.

QUESTION 21

Answer: D

Explanation: When using a balance, the correct procedure is to read the mass from the position of the pointer on the balance scale once it has stabilized. This ensures an accurate measurement of the object's mass.

QUESTION 22

Answer: C

Explanation: Scientific notation is used to express large or small numbers more conveniently. In Earth and space science, large distances such as the diameter of a planet, distances between celestial bodies, or the size of galaxies are often expressed using scientific notation.

Question 23

Answer: D

Explanation: Properly labeling chemical containers is essential for identifying the contents, potential hazards, and necessary safety precautions. It helps avoid accidental mix-ups and ensures safe handling and storage of chemicals in the laboratory.

QUESTION 24

Answer: A

Explanation: The density of a substance is typically expressed in grams per liter (g/L) in Earth and space science investigations. It represents the mass of a substance per unit volume and is commonly used to compare the densities of different materials.

QUESTION 25

Answer: D

Explanation: Safety should always be a top priority during outdoor Earth and space science investigations. Knowing the location of emergency exits and nearby shelters is crucial in case of sudden weather changes or unexpected emergencies. This helps ensure the well-being of both students and teachers during outdoor activities.

QUESTION 26

Answer: B

Explanation: Earth's axial tilt is the main factor responsible for the changing seasons. As the Earth orbits the Sun, the tilt of its axis remains relatively constant, causing different parts of the Earth to receive varying amounts of sunlight throughout the year, resulting in the four seasons.

QUESTION 27

Answer: C

Explanation: Igneous rocks are formed when lava or magma cools and solidifies. The cooling process can occur either on the Earth's surface (extrusive) or beneath the surface (intrusive). Examples of igneous rocks include basalt, granite, and pumice.

QUESTION 28

Answer: C

Explanation: Weathering is the process by which rocks are broken down into smaller particles through physical or chemical means. Physical weathering involves mechanical forces like temperature changes, frost action, and abrasion, while chemical weathering involves chemical reactions that alter the rock's composition.

QUESTION 29

Answer: C

Explanation: The ozone layer is primarily located in the stratosphere, which is the second layer of the Earth's atmosphere. The ozone layer plays a crucial role in absorbing harmful ultraviolet (UV) radiation from the Sun.

QUESTION 30

Answer: B

Explanation: Convergent plate boundaries occur when two tectonic plates move toward each other. In such regions, one plate is often forced beneath the other in a process called subduction. This subduction leads to the formation of volcanoes and can also trigger earthquakes as the plates interact and release built-up stress.

QUESTION 31

Answer: C

Explanation: Plate tectonics theory was developed based on the accumulation of evidence and ideas like continental drift proposed by Alfred Wegener, the discovery of seafloor spreading by Harry Hess, and the understanding of subduction zones. Options a, b, and d include scientific ideas that were either unrelated to plate tectonics or have been proven incorrect.

QUESTION 32

Answer: C

Explanation: The heliocentric model, proposed by Nicolaus Copernicus, shifted the understanding of the Universe from the geocentric model (Earth-centered) to the Sun-centered model. The heliocentric model laid the foundation for modern astronomy and cosmology. The other options, such as the steady-state theory and Ptolemaic system, are outdated and have been replaced by newer scientific ideas.

QUESTION 33

Answer: C

Explanation: Quantum mechanics is a major contemporary theory that provides the framework for understanding the behavior of matter at the atomic and subatomic levels. It explains phenomena such as wave-particle duality, quantization of energy levels, and the behavior of particles. The other options are outdated theories that have been replaced by modern scientific understanding.

QUESTION 34

Answer: D

Explanation: The theory of evolution, proposed by Charles Darwin, is a unifying theme that connects various branches of science, including biology, paleontology, genetics, and ecology. It also has implications for other STEM fields, such as medicine and agriculture. The other options are specific scientific theories or concepts that do not serve as broad unifying themes.

QUESTION 35

Answer: D

Explanation: Empiricism is a fundamental characteristic of the nature of science as a system of inquiry. It emphasizes the reliance on empirical evidence and observation to formulate and test scientific theories. The other options, such as dogmatism, fixed methodology, and subjectivity, do not accurately reflect the principles of the nature of science.

QUESTION 36

Answer: D

Explanation: The theory of natural selection, proposed by Charles Darwin and Alfred Russel Wallace, is central to the understanding of the origin and evolution of species. It explains how species adapt and evolve over time in response to changes in their environment. The other options are important scientific concepts, but they do not specifically address the origin and evolution of species.

QUESTION 37

Answer: A

Explanation: Catastrophism, an early geological theory, suggested that major geological changes were the result of catastrophic events like floods and earthquakes. This idea was later challenged by the concept of uniformitarianism, which proposed that geological processes occur gradually and are consistent over time. The other options are unrelated to the Earth's geological processes and the concept of uniformitarianism.

QUESTION 38

Answer: D

Explanation: The theory of genetics, pioneered by Gregor Mendel and further developed through modern molecular biology, explains how genetic material is passed from one generation to another. This principle connects the fields of chemistry and biology, as genetics is based on the study of genes and their chemical nature. The other options are unrelated to the continuity of life through genetic material.

QUESTION 39

Answer: C

Explanation: One key limitation of scientific models is that they cannot predict future events with absolute certainty. Models are simplified representations of complex natural phenomena, and their predictions are based on available data and assumptions. However, uncertainties, unforeseen variables, and the dynamic nature of natural systems can limit the accuracy of predictions. The other options do not accurately reflect the key limitation of scientific models.

QUESTION 40

Answer: C

Explanation: Tentativeness is a characteristic of the nature of science that acknowledges the provisional nature of scientific knowledge. Scientific ideas are subject to change and refinement as new evidence and data emerge. This feature encourages scientists to continuously reassess and revise their theories to align with the most current understanding of the natural world. The other options suggest fixed or unchangeable aspects of science, which is not in line with its tentativeness.

QUESTION 41

Answer: D

Explanation: Weathering is the process by which rocks and minerals are broken down into smaller particles at or near the Earth's surface. It can occur through physical processes (like frost action and abrasion), chemical processes (such as dissolution and oxidation), and biological processes (including the actions of plants and organisms). Seafloor spreading and plate tectonics are related to the movement of Earth's lithospheric plates, while rock magnetism pertains to the study of magnetic properties in rocks.

QUESTION 42

Answer: C

Explanation: Fossil fuel reservoirs, such as coal, oil, and natural gas deposits, are formed by the accumulation and compaction of organic material from ancient plants and organisms over millions of years. Volcanic craters are formed by volcanic activity, deltas are landforms shaped by the deposition of sediment at the mouth of a river, and sinkholes are created due to the collapse of underground caverns.

QUESTION 43

Answer: B

Explanation: Radioactive decay, the breakdown of radioactive isotopes in Earth's mantle and crust, is the primary source of energy that drives Earth's internal processes. This energy contributes to plate tectonics, which results in the movement of lithospheric plates and leads to volcanic activity, earthquakes, and mountain formation. Solar radiation is the primary source of energy for Earth's surface processes, but it is not directly responsible for internal geological processes.

QUESTION 44

Answer: C

Explanation: Bernoulli's principle states that as the velocity of a fluid increases, the pressure exerted by that fluid decreases, and vice versa. This principle is crucial in understanding various atmospheric and oceanic phenomena, including airflow around mountains, flight dynamics, and weather patterns. Boyle's law relates the pressure and volume of a gas at a constant temperature, Archimedes' principle describes buoyancy, and Pascal's principle is concerned with pressure transmission in confined fluids.

QUESTION 45

Answer: B

Explanation: Glacial isostasy is a major Earth science concept that explains the vertical movement of the Earth's crust in response to the formation and melting of glaciers. When glaciers form, they exert immense pressure on the Earth's surface, causing the land beneath to depress or subside. As glaciers melt, this pressure is released, leading to the rebound or uplift of the land. Seafloor spreading is related to the movement of tectonic plates, the rock cycle describes the transformation of rocks, and erosion involves the wearing away and transport of materials by natural agents like water, wind, or ice.

QUESTION 46

Answer: A

Explanation: Nicolaus Copernicus, a Polish astronomer and mathematician, developed the heliocentric model in the 16th century, challenging the geocentric view prevailing at the time.

QUESTION 47

Answer: B

Explanation: Solar energy from the Sun is the primary driver of Earth's climate and weather systems. It influences temperature, atmospheric circulation, and various weather phenomena.

QUESTION 48

Answer: A

Explanation: Nitrogen fixation is the process by which certain bacteria convert atmospheric nitrogen (N_2) into ammonia (NH_3) or other nitrogen compounds that can be used by plants for growth.

QUESTION 49

Answer: B

Explanation: Plate Tectonics is the scientific theory that describes the movement of Earth's lithospheric plates and their interactions, leading to the formation of mountains, earthquakes, volcanoes, and other geological features.

QUESTION 50

Answer: C

Explanation: Igneous rock is formed from the cooling and solidification of molten lava or magma. It can be found both on the Earth's surface and beneath it.

QUESTION 51

Answer: A

Explanation: The greenhouse effect is primarily caused by the accumulation of greenhouse gases in the atmosphere, with carbon dioxide being the most significant one. These gases trap heat and warm the planet.

QUESTION 52

Answer: C

Explanation: Chemistry is the branch of science that deals with the study of matter, its composition, structure, and properties, as well as the changes it undergoes during chemical reactions.

QUESTION 53

Answer: A

Explanation: Scientific notation is often used to express large numbers or astronomical time scales. "10^7 years" represents ten million years, which is relevant when discussing the age of Earth.

QUESTION 54

Answer: C

Explanation: Archimedes' Principle states that an object submerged in a fluid experiences an upward buoyant force equal to the weight of the fluid it displaces. This principle is essential for understanding buoyancy.

QUESTION 55

Answer: D

Explanation: Formulating a hypothesis is a step in the scientific method where researchers make predictions or propose a tentative explanation for the observed phenomenon. Subsequent steps involve testing the hypothesis through experimentation, data analysis, and drawing conclusions based on the results.

QUESTION 56

Answer: D

Explanation: The Theory of Paleomagnetism explains the study of Earth's ancient magnetic field recorded in rocks. It provides evidence for the phenomenon of magnetic pole reversals in Earth's history.

QUESTION 57

Answer: C

Explanation: Sedimentary rocks are formed from the accumulation and compression of sediments, which may include mineral fragments, organic matter, and other debris deposited in water bodies over time.

QUESTION 58

Answer: C

Explanation: Milankovitch cycles refer to the variations in Earth's orbital eccentricity, axial tilt, and precession that influence the amount and distribution of solar radiation reaching Earth, leading to long-term climate changes.

QUESTION 59

Answer: B

Explanation: The Richter scale is a logarithmic scale used to measure the energy released by earthquakes. It quantifies the seismic magnitude and the ground motion associated with an earthquake.

QUESTION 60

Answer: C

Explanation: Shield volcanoes are large, broad volcanoes with gentle slopes, resembling a warrior's shield lying on the ground. They are primarily formed by the accumulation of low-viscosity basaltic lava flows during non-explosive eruptions.

QUESTION 61

Answer: D

Explanation: Earth and space sciences focus on understanding the Earth and its natural processes, including its geology, atmosphere, and space environment. Maps, models, and geospatial technologies are essential tools used by earth science teachers to present scientific information effectively.

Question 62

Answer: D

Explanation: Earth and space science is the field of study that explores the interrelationships of science, technology, engineering, and mathematics in understanding the Earth, its processes, and the space environment.

QUESTION 63

Answer: C

Explanation: Evaluating scientific research and media coverage helps earth science teachers ensure that accurate and reliable scientific information is conveyed to the public and students, promoting a better understanding of Earth and space sciences.

QUESTION 64

Answer: C

Explanation: Technology and society is the field of study that delves into the social, economic, and ethical issues associated with technological and scientific developments, including those relevant to Earth and space sciences.

QUESTION 65

Answer: D

Explanation: Maps, models, and geospatial technologies are tools that help earth science teachers present scientific information in a clear and visually engaging manner, aiding in the understanding of Earth and space sciences.

QUESTION 66

Answer: D

Explanation: Analyzing social, economic, and ethical issues in Earth and space sciences involves investigating the potential implications and consequences of scientific and technological advancements.

QUESTION 67

Answer: C

Explanation: While mathematics plays a significant role in scientific research and data analysis, it is not considered a separate branch of Earth and space sciences. The other options, geology, astronomy, and atmospheric science, are all integral parts of the field.

QUESTION 68

Answer: C

Explanation: Evaluating scientific research and media coverage in Earth and space sciences is done to ensure that the information presented is accurate, reliable, and supported by scientific evidence.

QUESTION 69

Answer: C

Explanation: Scientific visualization focuses on understanding how scientific information is visually represented using tools such as maps, models, and geospatial technologies in the context of Earth and space sciences.

QUESTION 70

Answer: C

Explanation: Earth and space sciences encompass the study of Earth, its natural processes, geology, atmosphere, and the space environment. The other options are related to specific aspects within the broader field of Earth and space sciences, but they do not fully encompass the entire field.

QUESTION 71

Answer: C

Explanation: Relative dating relies on the principle of superposition, where older rocks are found below younger ones.

QUESTION 72

Answer: B

Explanation: The geologic time scale uses fossils found in rocks to identify and classify various periods in Earth's history.

QUESTION 73

Answer: C

Explanation: Radiometric dating is based on the decay of radioactive isotopes to determine the exact age of rocks.

QUESTION 74

Answer: A

Explanation: The K-Pg boundary is a significant geological event marked by a mass extinction event, including the dinosaurs.

QUESTION 75

Answer: C

Explanation: Volcanic outgassing released gases, including water vapor and carbon dioxide, contributing to Earth's early atmosphere and hydrosphere.

QUESTION 76

Answer: A

Explanation: The nebular hypothesis proposes that the solar system formed from a rotating cloud of gas and dust.

QUESTION 77

Answer: B

Explanation: Petrification occurs when organic materials are replaced by minerals, preserving the remains as fossils.

QUESTION 78

Answer: B

Explanation: Fossil records show the development and changes in species over geological time, supporting the theory of evolution.

QUESTION 79

Answer: C

Explanation: The Mesozoic Era is often referred to as the "Age of Reptiles" because of the prevalence of dinosaurs.

QUESTION 80

Answer: D

Explanation: Biological evolution refers to the change in the characteristics of living organisms over successive generations.

QUESTION 81

Answer: C

Explanation: Nitrogen was the primary gas in Earth's early atmosphere, along with smaller amounts of carbon dioxide and water vapor.

QUESTION 82

Answer: B

Explanation: Paleomagnetic dating relies on the recording of the Earth's past magnetic field in rocks.

QUESTION 83

Answer: D

Explanation: The theory of plate tectonics suggests that Earth's lithosphere is fragmented into several tectonic plates.

QUESTION 84

Answer: C

Explanation: Prokaryotes, such as bacteria, were among the first life forms to appear on Earth.

QUESTION 85

Answer: A

Explanation: Amber is the fossilized resin of ancient trees, often preserving organisms trapped within it.

QUESTION 86

Answer: A

Explanation: Option A is the correct answer. The fossil discovery, exhibiting the same species on two separate continents, aligns with the concept of continental drift. It suggests that the continents were once connected, allowing for migration and the spread of species. To support the theory further, geologists can use additional lines of evidence such as matching rock formations, glacial striations, and the fit of continental coastlines. Conversely, skeptics could cite the possibility of land bridges, sea currents, or other geological processes to challenge the connection between the fossil and continental drift.

QUESTION 87

Answer: D

Explanation: Option D is the correct answer. The asteroid impact hypothesis is backed by various pieces of evidence, including the identification of impact craters (e.g., Chicxulub crater), the presence of shocked quartz in sedimentary layers, and abnormal iridium concentrations. While volcanic eruptions and climate change can also cause mass extinctions, the specific evidence for an asteroid impact strengthens the case for this particular scenario.

QUESTION 88

A group of researchers claims to have found the earliest known evidence of life on Earth, suggesting the existence of microbial

Answer: B

Explanation: Option B is the correct answer. While the researchers' discovery of potential microbial mats is interesting, it is crucial to consider alternative explanations. Some structures resembling microbial mats can arise through abiotic processes or mineral formations, leading to misinterpretations. Further investigations, such as microscopic analysis and detailed mineralogical studies, would be necessary to validate or refute their claims.

QUESTION 89

Answer: B

Explanation: Option B is the correct answer. Adaptive radiation is the process by which a single ancestral species diversifies into a variety of different species, each adapted to occupy specific niches or habitats. The Galapagos finches are a classic example, where different beak shapes evolved to exploit distinct food sources on different islands. Environmental factors, such as available food types and competition, played a crucial role in facilitating these divergent adaptations.

QUESTION 90

Answer: B

Explanation: Option B is the correct answer. Archaeopteryx and other transitional fossils display a combination of dinosaur and bird traits, indicating an evolutionary link between these two groups. The presence of feathers, a characteristic primarily associated with birds, in a dinosaur-like body provides compelling evidence for the gradual transition from dinosaurs to birds. Additionally, skeletal features, such as a wishbone and hollow bones, support the evolutionary relationship. This evidence strengthens the theory of evolution and helps elucidate the origin of birds from dinosaurs.

QUESTION 91

Answer: B

Explanation: The primary driving force behind tectonic plate movement is convection currents in the Earth's mantle. Heat from the Earth's core creates convection currents, causing the semi-fluid asthenosphere to circulate. This movement of the mantle material drags the tectonic plates, leading to their motion on the Earth's surface.

QUESTION 92

Answer: C

Explanation: Stratovolcanoes, also known as composite volcanoes, are characterized by highly explosive eruptions and steep-sided slopes. These volcanoes are built up by alternating layers of lava flows, volcanic ash, and other volcanic materials. The eruptions of stratovolcanoes are often violent due to the buildup of thick, viscous lava that can block the vent, leading to pressure buildup and explosive eruptions.

QUESTION 93

Answer: B

Explanation: Earthquakes are caused primarily by the sliding of tectonic plates past each other along faults. These faults are fractures in the Earth's crust where the rocks on either side have moved relative to each other. The stress and strain buildup along these faults lead to sudden releases of energy, resulting in seismic waves and earthquakes.

QUESTION 94

Answer: B

Explanation: Metamorphism is the process by which existing rocks are subjected to high temperatures and pressures within the Earth's crust, leading to their transformation into new rocks. This process occurs without melting the rock entirely. Metamorphic rocks form due to the recrystallization of minerals, resulting in changes in texture and mineral composition while the overall chemical composition of the rock may remain relatively unchanged.

QUESTION 95

Answer: C

Explanation: Basalt is an igneous rock that forms from rapid cooling of lava at or near the Earth's surface. The quick cooling prevents large mineral crystals from forming, resulting in a fine-grained texture. Basalt is one of the most common types of volcanic rock and is found in various volcanic terrains and oceanic crust.

QUESTION 96

Answer: C

Explanation: Sedimentary rocks are formed from the accumulation and cementation of mineral and organic particles (sediments) in various environments such as riverbeds, lakes, oceans, and deserts. Over time, these sediments are compacted and cemented together to form sedimentary rocks. Examples of sedimentary rocks include sandstone, limestone, shale, and conglomerate.

QUESTION 97

Answer: A

Explanation: Mid-ocean ridges are associated with divergent plate boundaries. At divergent boundaries, tectonic plates move away from each other, and new oceanic crust is formed as magma rises from the mantle and solidifies at the mid-ocean ridge. This process is responsible for the continuous expansion of the ocean floor and the creation of new crust.

QUESTION 98

Answer: B

Explanation: Paleomagnetic data is one of the key lines of evidence used to support the theory of plate tectonics. Rocks contain magnetic minerals that align with the Earth's magnetic field at the time of their formation. By studying the orientation of these magnetic minerals in ancient rocks, geologists can determine the past positions and movements of the Earth's tectonic plates, providing strong evidence for plate tectonics.

QUESTION 99

Answer: A

Explanation: A caldera is a large, steep-walled depression that forms after a massive volcanic eruption, where the magma chamber is emptied, and the volcano collapses into the void. These volcanic features can be several kilometers in diameter and are often associated with highly explosive eruptions. Crater Lake in Oregon, USA, is an example of a caldera formed by the collapse of Mount Mazama after a volcanic eruption.

QUESTION 100

Answer: B

Explanation: Erosion is the process responsible for the gradual wearing away of Earth's surface through the action of various agents such as wind, water, ice, and gravity. These agents transport and remove weathered materials from one location to another, shaping the landscape over time. Erosion is a key component of the rock cycle, as it contributes to the breakdown and transportation of rock fragments that can eventually become part of sedimentary rocks.

QUESTION 101

Answer: A

Explanation: A normal fault is characterized by the movement of rock blocks away from each other due to tensional forces. In this type of fault, the hanging wall (the block above the fault) moves downward relative to the footwall (the block below the fault). Normal faults are often associated with regions experiencing extension and are common in areas undergoing rifting or

QUESTION 102

Answer: C

Explanation: One of the main challenges related to the theory of plate tectonics is explaining the distribution of ancient fossils across continents. The theory suggests that continents were once part of larger supercontinents (e.g., Pangaea) and have since drifted apart. However, the presence of identical or similar fossils on continents that are now widely separated by oceans raises questions about how these organisms could have migrated across vast distances of ocean.

QUESTION 103

Answer: C

Explanation: In this example, the rock cycle starts with volcanic activity, where molten magma from the Earth's mantle erupts onto the surface and cools rapidly, forming an igneous rock. Over time, weathering and erosion processes break down the igneous rock into sediments that are transported by wind, water, or ice and eventually deposited in a basin. The accumulation and compaction of these sediments lead to the formation of sedimentary rock.

QUESTION 104

Answer: C

Explanation: Each of the three measures has a significant role in earthquake preparedness:

Developing early warning systems: Early warning systems can detect the initial seismic waves before the more damaging waves arrive, giving people a few seconds to minutes of warning. This can provide valuable time to take cover, evacuate hazardous areas, or shut down critical infrastructure and processes, reducing the potential for injuries and damage.

Practicing "Drop, Cover, and Hold On" drills: These drills teach individuals the appropriate actions to take during an earthquake. Dropping to the ground prevents being knocked over, taking cover under furniture or a sturdy object protects against falling debris, and holding on provides stability during shaking. Regular practice of these drills helps ensure that people react quickly and correctly when an earthquake occurs.

Reinforcing critical infrastructure: Strengthening critical infrastructure, such as bridges, schools, hospitals, and utility systems, can mitigate the extent of damage during an earthquake. Proper design and retrofitting can enhance the structural integrity of buildings and infrastructure, making them more resistant to seismic forces and reducing the risk of collapse.

QUESTION 105

Answer: A

Explanation: The Cretaceous-Paleogene (K-Pg) mass extinction, also known as the Cretaceous-Tertiary (K-T) extinction, is one of the most well-known mass extinctions in Earth's history. It occurred approximately 66 million years ago and marks the end of the Cretaceous period and the beginning of the Paleogene period.

The prevailing theories regarding its cause are related to both volcanic activity and an asteroid impact:

Volcanic Activity: Some scientists propose that large-scale volcanic eruptions, particularly the eruption of the Deccan Traps in present-day India, played a significant role in the K-Pg mass extinction. These volcanic events released massive amounts of greenhouse gases into the atmosphere, leading to global climate change, acid rain, and disruption of the marine and terrestrial ecosystems.

Asteroid Impact: The most widely accepted theory is that a large asteroid or comet impact was the primary cause of the K-Pg mass extinction. This hypothesis is supported by a layer of sediment found worldwide, known as the K-Pg boundary, containing

QUESTION 106

Answer: C

Explanation: Quartz and feldspar are two common examples of minerals found in the Earth's crust. Minerals are naturally occurring, inorganic solid substances with a definite chemical composition and a specific internal crystal structure.

QUESTION 107

Answer: B

Explanation: Mining is the process of extracting valuable minerals or other geological materials from the Earth's crust. It is an essential activity for obtaining various natural resources and materials used in industries and daily life.

QUESTION 108

Answer: C

Explanation: Igneous rocks are formed from the cooling and solidification of magma (below the Earth's surface) or lava (on the Earth's surface). This process leads to the formation of rocks with a crystalline structure, such as basalt and granite.

QUESTION 109

Answer: B

Explanation: Seismology is the study of seismic waves, which are generated by earthquakes or other seismic sources, and how they travel through and interact with the Earth's interior. It provides valuable information about the Earth's internal structure.

QUESTION 110

Answer: C

Explanation: Fossil records provide evidence of past life forms on Earth but do not directly offer insights into the Earth's internal structure. Seismic wave behavior, magnetic field measurements, and rock samples from drilling are used to study the Earth's interior.

QUESTION 111

Answer: A

Explanation: Weathering is the process by which rocks are broken down into smaller particles through physical and chemical processes. This can be caused by factors such as water, wind, temperature changes, and the actions of living organisms.

QUESTION 112

Answer: A

Explanation: Loam is a type of soil that contains roughly equal proportions of sand, silt, and clay. It is considered one of the best types of soil for plant growth because it provides a good balance of drainage and water retention.

QUESTION 113

Answer: C

Explanation: Natural gas is a non-renewable geologic resource. It is a fossil fuel formed from the remains of ancient plants and animals, and once it is depleted, it cannot be replenished within a human lifetime.

QUESTION 114

Answer: C

Explanation: The outer core of the Earth is primarily composed of molten iron and nickel. It is a liquid layer that lies beneath the mantle and surrounds the solid inner core.

QUESTION 115

Answer: A

Explanation: Subduction is the process by which one tectonic plate moves under another tectonic plate and sinks into the Earth's mantle. This occurs at convergent plate boundaries and leads to the recycling of crustal material.

QUESTION 116

Answer: C

Explanation: Metamorphic rocks form from pre-existing rock that has been altered by heat, pressure, or chemically active fluids without melting. Examples include marble (from limestone) and schist (from shale).

QUESTION 117

Answer: B

Explanation: Diamond is the hardest known naturally occurring substance on Earth. It has a Mohs hardness of 10, which means it cannot be scratched by any other mineral.

QUESTION 118

Answer: D

Explanation: The O horizon is the topmost soil horizon and contains organic matter in various stages of decomposition, such as leaves, twigs, and other plant material. It is commonly known as the organic layer.

QUESTION 119

Answer: A

Explanation: Coal is a geologic resource formed from the remains of plants and animals that lived millions of years ago and have been subjected to heat and pressure over time. It is a combustible black or brownish-black sedimentary rock.

QUESTION 120

Answer: C

Explanation: Seismology is the method commonly used to determine the internal structure of the Earth. By analyzing how seismic waves travel through the planet during earthquakes, scientists can infer the properties of the Earth's interior.

QUESTION 121

Answer: C

Explanation: During a major volcanic eruption in a densely populated area, the immediate effects can be catastrophic. The eruption may release toxic gases, such as sulfur dioxide and carbon dioxide, which can cause respiratory problems and contribute to air pollution. Ash and pyroclastic flows, which are fast-moving currents of hot gas and volcanic material, can devastate everything in their path, leading to loss of life, destruction of property, and disruption of communities. These immediate effects can have severe consequences for human health, agriculture, and the environment.

QUESTION 122

Answer: C

Explanation: Plate tectonics is a scientific theory that explains the movement and interaction of the Earth's lithosphere (the rigid outer layer of the Earth) on the semi-fluid asthenosphere (a partially molten layer beneath the lithosphere). According to this theory, the lithosphere is divided into several tectonic plates that are in constant motion, drifting and colliding with each other. These movements lead to various geological phenomena such as earthquakes, volcanic activity, mountain formation, and the opening and closing of ocean basins.

QUESTION 123

Answer: C

Explanation: Soil is a vital natural resource that supports life on Earth. It is a complex mixture of mineral particles (sand, silt, and clay), organic matter (decaying plant and animal material), water, air, and living organisms like bacteria, fungi, and small insects. This rich composition makes soil an essential component for supporting plant growth as it provides nutrients, water, and a stable physical structure for roots. Soil also serves as a habitat for countless organisms, from microorganisms to larger animals, contributing to the overall biodiversity of ecosystems.

QUESTION 124

Answer: A

Explanation: The key difference between rocks and minerals lies in their formation and composition. Minerals are naturally occurring, inorganic substances with a specific chemical composition and a regular internal crystal structure. They can be formed through various geological processes, including the cooling and solidification of molten rock (as in the case of igneous minerals) or through precipitation from a solution (as in the case of some sedimentary minerals). On the other hand, rocks are aggregates of one or more minerals and/or mineraloids. They can be formed through different processes, such as the cooling and solidification of magma or lava to create igneous rocks, the compaction and cementation of sediments to form sedimentary rocks, or the alteration of existing rocks through heat and pressure to create metamorphic rocks.

QUESTION 125

Answer: C

Explanation: Climate change is a complex phenomenon influenced by both natural processes and human activities. While natural processes do contribute to changes in the Earth's climate over geological timescales, human activities have accelerated the rate of climate change in recent times. The burning of fossil fuels (such as coal, oil, and natural gas) for energy, deforestation (which reduces the capacity of plants to absorb

QUESTION 126

Answer: A

Explanation: The Law of Superposition states that in an undisturbed rock sequence, the oldest rocks are found at the bottom, and each successive layer is younger than the one beneath it. This principle is fundamental in determining the relative ages of rock layers.

QUESTION 127

Answer: A

Explanation: Index fossils are species that existed for relatively short periods of time but were widespread geographically. They are excellent markers for correlating and dating rock layers across large distances, as they help identify specific geological time periods.

QUESTION 128

Answer: A

Explanation: Radioactive decay is the process in which an unstable atomic nucleus loses energy by emitting radiation, such as alpha particles, beta particles, or gamma rays. This process is essential for understanding the age of rocks and the isotopic composition of elements.

QUESTION 129

Answer: B

Explanation: At a divergent plate boundary, two tectonic plates move away from each other. This movement allows magma to rise from the mantle, creating new crust and contributing to seafloor spreading.

QUESTION 130

Answer: A

Explanation: Subduction occurs at convergent plate boundaries when one tectonic plate descends beneath another and sinks into the mantle. This process often leads to the formation of deep ocean trenches and volcanic arcs.

QUESTION 131

Answer: C

Explanation: Igneous rocks form from the cooling and solidification of magma (below the Earth's surface) or lava (on the Earth's surface). The process of solidification can result in various textures, such as fine-grained or coarse-grained, depending on the cooling rate.

QUESTION 132

Answer: D

Explanation: The lithosphere is not a compositional layer but rather a mechanical layer of the Earth. It consists of the rigid outermost part of the Earth, including the crust and the uppermost part of the mantle.

QUESTION 133

Answer: D

Explanation: Minerals are naturally occurring, inorganic substances with a specific chemical composition and an ordered internal crystal structure. They are the building blocks of rocks and play a crucial role in various geological processes.

QUESTION 134

Answer: B

Explanation: Resource management refers to the sustainable and responsible utilization of Earth's materials and resources to fulfill present and future needs. It involves careful planning, conservation efforts, and consideration of environmental impacts.

QUESTION 135

Answer: C

Explanation: Mechanical weathering, also known as physical weathering, involves the physical breakdown of rocks into smaller particles without altering their chemical composition. Examples include frost wedging, abrasion, and exfoliation.

QUESTION 136

Answer: C

Explanation: Deposition is the process by which sediments, soil, and rock fragments are laid down or accumulated in new locations. This occurs after the materials have been eroded and transported by various agents like water, wind, or ice.

QUESTION 137

Answer: C

Explanation: A cirque is a bowl-shaped hollow with steep walls, typically found in mountainous regions and formed by glacial erosion. It is the initial stage in the formation of a glacier.

QUESTION 138

Answer: A

Explanation: The presence of marine fossils in the shale layer indicates a marine environment during its deposition. The unconformity between the shale and sandstone layers suggests a period of non-deposition or erosion, which was followed by the deposition of the terrestrial sandstone layer. This sequence of rocks and the unconformity provide valuable evidence of a transition from a marine environment to a terrestrial environment over time.

QUESTION 139

Answer: C

Explanation: The similarity in volcanic rocks and marine fossils on both islands suggests that they were once connected. Through the process of plate tectonics, they drifted apart, leading to their current separation. As Island A moved away from its original location, it experienced subsidence (sinking) due to the weight of accumulated sediments, causing the volcanic rocks to become progressively older as new layers formed.

QUESTION 140

Answer: A

Explanation: Assessing the economic viability and environmental impacts of a mining project requires a comprehensive analysis. The physical properties and chemical composition of the mineral are crucial in determining its market demand and economic value. Additionally, evaluating the accessibility of the site and potential transportation costs are essential for calculating the project's overall profitability. The potential environmental impacts of mining activities, including habitat disruption, water pollution, and community displacement, must be thoroughly evaluated to ensure responsible and sustainable mining practices.

QUESTION 141

Answer: A

Explanation: Physical weathering refers to the mechanical breakdown of rocks into smaller pieces without any change in their chemical composition. This process includes actions like frost wedging, abrasion, and root wedging. On the other hand, chemical weathering involves the alteration of the chemical structure of rocks through chemical reactions with substances like water, acids, or gases. Both processes play essential roles in shaping the Earth's surface.

QUESTION 142

Answer: C

Explanation: U-shaped valleys are characteristic features of landscapes shaped by glaciation. Alpine glaciers are responsible for carving out these valleys through a combination of abrasion and plucking. As glaciers move down mountains, they scrape and erode the surrounding rock, forming U-shaped valleys over long periods of time.

QUESTION 143

Answer: C

Explanation: Deposition is the process of dropping or settling of sediments by agents like water, wind, or ice. During transportation, water moves sediments and sorts them based on size, with larger and heavier particles settling first and finer particles transported further. This process is essential in the formation of sedimentary rocks and shaping various landforms.

QUESTION 144

Answer: A

Explanation: Chemical weathering can dissolve and alter certain types of rocks, leading to the formation of caves and sinkholes. Over time, the chemical action of water can create underground cavities and passages in limestone and other soluble rocks, resulting in the formation of caves. Sinkholes are depressions on the Earth's surface caused by the collapse of caves or dissolution of bedrock.

QUESTION 145

Answer: D

Explanation: Volcanic eruptions are not considered erosional processes. Instead, they are geological events where molten rock, ash, and gases are expelled from a volcano. Erosional processes, on the other hand, involve the movement and removal of soil, rock, or sediment by agents like wind, water, ice, and gravity.

QUESTION 146

Answer: A

Explanation: Fjords are deep, narrow inlets with steep cliffs on either side and are typically found in coastal areas. These landforms are a result of the past action of alpine glaciers, which carved out these valleys during periods of glaciation. As the glaciers retreated, seawater flooded the valleys, forming fjords.

QUESTION 147

Answer: C

Explanation: The size of sediment particles influences their settling and transport in water. Larger and heavier particles tend to settle more slowly, while smaller and lighter particles are carried further downstream before settling. As a result, larger particles are often deposited closer to the source of the sediment, while smaller particles may travel longer distances before settling.

QUESTION 148

Answer: A

Explanation: Temperature and humidity play significant roles in chemical weathering processes. Higher temperatures and moisture promote chemical reactions that break down minerals in rocks more quickly. In contrast, regions with lower temperatures and arid conditions may experience slower rates of chemical weathering.

QUESTION 149

Answer: A

Explanation: Weathering is the process of breaking down rocks into smaller fragments through mechanical or chemical means, but it does not involve the movement of these fragments. Erosion, on the other hand, is the transportation and movement of the weathered sediments by agents such as water, wind, ice, or gravity.

QUESTION 150

Answer: B

Explanation: A piedmont glacier forms when a valley glacier flows out of the mountains and spreads into a broad lobe-like shape in a lowland or coastal plain. As the glacier exits the confining valley, it expands horizontally, creating a wide, fan-shaped landform.

QUESTION 151

Answer: C

Explanation: Deforestation involves the removal of trees and vegetation from an area, leaving the soil exposed and vulnerable to erosion. The roots of plants play a crucial role in stabilizing the soil and preventing erosion. When trees are cut down and vegetation is removed, rainwater can wash away the topsoil more easily, leading to increased rates of erosion.

QUESTION 152

Answer: B

Explanation: Sand dunes are formed through the process of wind erosion and deposition. Wind erodes loose particles of sand from the desert floor and carries them through the air. When the wind slows down or encounters an obstacle, the sand is deposited, creating dunes of various shapes and sizes.

QUESTION 153

Answer: C

Explanation: The size of the glacier has a significant impact on the rate of glacial erosion. Larger glaciers, with more ice and greater mass, can exert more pressure on the underlying rock and move more debris, resulting in higher rates of erosion compared to smaller glaciers.

QUESTION 154

Answer: C

Explanation: Climate and geography are essential factors influencing the formation of various landforms on Earth. For example, arid climates lead to the formation of deserts, tectonic activity shapes mountains, and rivers carve out valleys as they flow through different geological regions. These variations contribute to the diversity of landscapes we observe on our planet.

QUESTION 155

Answer: A

Explanation: The hydrologic cycle begins with evaporation, where water changes from liquid to vapor, followed by condensation, where water vapor turns back into liquid, then precipitation, where water falls back to the Earth's surface, and finally infiltration, where water seeps into the ground.

QUESTION 156

Answer: C

Explanation: The primary energy source driving the water cycle is solar energy, which heats the Earth's surface, leading to evaporation of water bodies and causing water to enter the atmosphere. Solar energy is a form of thermal energy.

QUESTION 157

Answer: B

Explanation: Condensation is the process where water vapor in the atmosphere cools and transforms back into liquid water droplets, which then accumulate to form clouds.

QUESTION 158

Answer: D

Explanation: pH level is a chemical property of water that indicates its acidity or alkalinity. It is a measure of the concentration of hydrogen ions in the water.

QUESTION 159

Answer: B

Explanation: During the hydrologic cycle, as water evaporates from the ocean or other bodies of water, it leaves behind the dissolved salts, leading to an increase in salinity. When the water condenses to form clouds and eventually precipitates, it is relatively pure, and the salinity decreases. As water infiltrates into the ground, it may pick up minerals and nutrients, but its salinity decreases compared to the initial precipitation.

QUESTION 160

Answer: D

Explanation: The cryosphere refers to the frozen components of the Earth, such as glaciers, ice caps, and permafrost. While it interacts with the hydrosphere through processes like ice melting and iceberg formation, it is not directly interconnected as the other options.

QUESTION 161

Answer: B

Explanation: Ocean currents play a crucial role in redistributing heat around the planet. Warm ocean currents carry heat from the equator towards the poles, while cold currents transport cold water from the poles towards the equator. This redistribution of heat significantly influences regional climates and weather patterns.

QUESTION 162

Answer: C

Explanation: Oceans and seas cover a significant portion of the Earth's surface and are the primary sources of water vapor in the atmosphere. Evaporation from these water bodies releases water vapor into the air, which contributes to the hydrologic cycle.

QUESTION 163

Answer: B

Explanation: A point source of water pollution is a specific, identifiable location where pollutants enter a water body, such as a river or a lake. Industrial discharge refers to the direct release of pollutants from industrial facilities, making it a classic example of point source pollution.

QUESTION 164

Answer: A

Explanation: Wetlands play a crucial role in the hydrosphere by acting as natural reservoirs for freshwater. They help regulate water flow, retain and slowly release water, which is vital for maintaining water availability during dry periods or floods.

QUESTION 165

Answer: A

Explanation: Sublimation is the process where a solid (like ice or snow) transforms directly into water vapor without passing through the liquid state. It is the opposite of deposition, where water vapor changes directly into ice without becoming a liquid.

QUESTION 166

Answer: A

Explanation: Human activities, particularly the burning of fossil fuels and deforestation, release significant amounts of carbon dioxide into the atmosphere. This leads to increased carbon dioxide dissolution in water bodies, changing the water's chemical properties and affecting its acidity.

QUESTION 167

Answer: A

Explanation: Transpiration is the process by which water is taken up by plant roots and released through small pores on leaves called stomata. This water vapor is then released into the atmosphere.

QUESTION 168

Answer: B

Explanation: Solar radiation from the Sun is the primary driving force that causes evaporation of water from oceans, lakes, and rivers, initiating the hydrologic cycle.

QUESTION 169

Answer: A

Explanation: The epipelagic zone (sunlight zone) is the uppermost layer of the ocean where photosynthesis occurs, leading to higher oxygen levels.

QUESTION 170

Answer: D

Explanation: Fjords are deep, narrow inlets formed by glacial erosion and are usually connected to the ocean, making them part of the marine ecosystem rather than a freshwater system.

QUESTION 171

Answer: A

Explanation: Sublimation is the phase transition in which water vapor transforms directly into ice without becoming a liquid.

QUESTION 172

Answer: C

Explanation: The Gulf Stream is a powerful, warm ocean current that flows along the eastern coast of North America, influencing the climate of the surrounding regions.

QUESTION 173

Answer: C

Explanation: Infiltration is the process by which water seeps into the soil and permeable rock layers, replenishing groundwater.

QUESTION 174

Answer: C

Explanation: Ocean tides are primarily caused by the gravitational forces of the Moon and the Sun, as well as the centrifugal force resulting from the Earth-Moon system's orbital motion.

QUESTION 175

Answer: C

Explanation: Braided rivers have multiple interconnected channels that divide and recombine, creating a braided pattern, often occurring in areas with high sediment load and variable flow rates.

QUESTION 176

Answer: A

Explanation: The major source of dissolved salts in seawater is the weathering of rocks on land and their subsequent transportation to the ocean through river runoff.

QUESTION 177

Answer: C

Explanation: Wetlands are areas of slow-moving or stagnant water, and many are covered with a layer of floating vegetation like lilies and cattails.

QUESTION 178

Answer: B

Explanation: Condensation is the process of water vapor transforming into liquid water when it cools and reaches its dew point.

QUESTION 179

Answer: C

Explanation: The Coriolis effect, caused by Earth's rotation, deflects moving air and water to the right in the Northern Hemisphere and to the left in the Southern Hemisphere, influencing ocean currents.

QUESTION 180

Answer: C

Explanation: Transpiration is the process by which water is lost from plants through their leaves in the form of water vapor.

QUESTION 181

Answer: B

Explanation: An aquifer is a layer of permeable rock or sediment that can store and transmit groundwater.

QUESTION 182

Answer: A

Explanation: Latitude plays a significant role in determining the salinity levels of different oceans, with higher salinity typically found in subtropical regions and lower salinity near the equator and polar regions.

QUESTION 183

Answer: C

Explanation: Precipitation is the main source of water in a watershed, and its distribution and intensity significantly influence stream flow.

QUESTION 184

Answer: C

Explanation: Surface water, such as rivers and lakes, is easily accessible and can be used directly for various human activities like drinking, irrigation, and industrial purposes.

QUESTION 185

Answer: D

Explanation: The permeability of the aquifer, which refers to how easily water can flow through it, greatly influences the movement and distribution of groundwater.

QUESTION 186

Answer: D

Explanation: The thickness of aquifers directly impacts the storage and availability of groundwater, making it a significant geological factor affecting freshwater resources.

QUESTION 187

Answer: A

Explanation: A watershed, also known as a drainage basin, is an area of land where all the water that falls or flows within it ultimately drains to a common point.

QUESTION 188

Answer: D

Explanation: Discharging untreated industrial waste into rivers can introduce pollutants, chemicals, and toxins, thereby negatively impacting the quality of surface water.

QUESTION 189

Answer: B

Explanation: In a confined aquifer, the primary mode of water movement is vertical flow, as the aquifer is sandwiched between two impermeable layers, causing water to move up or down.

QUESTION 190

Answer: B

Explanation: The soil composition, including its permeability and porosity, influences the rate at which water infiltrates into the ground, impacting groundwater recharge.

QUESTION 191

Answer: B

Explanation: A dendritic drainage pattern resembles the branching pattern of a tree, with smaller tributaries merging into larger rivers or streams.

QUESTION 192

Answer: C

Explanation: The discharge of a river is influenced by the velocity of the flowing water, as it determines the volume of water passing a given point per unit of time.

QUESTION 193

Answer: D

Explanation: The porosity of the rock refers to the amount of open space (pores) in it. Higher porosity increases the storage capacity of the aquifer for groundwater.

QUESTION 194

Answer: B

Explanation: Precipitation is the term used to describe the process of water falling from the atmosphere to the Earth's surface in the form of rain, snow, sleet, or hail.

QUESTION 195

Answer: C

Explanation: Over-pumping from wells can cause groundwater depletion, as the rate of water extraction exceeds the rate of natural recharge, leading to lowered water levels in aquifers.

QUESTION 196

Answer: B

Explanation: Runoff refers to the movement of water across the land surface, either overland flow or through channels, and occurs when the soil is saturated or impermeable.

QUESTION 197

Answer: C

Explanation: Surface water flow rate is not a direct factor that affects the quality of groundwater resources. The other options can introduce contaminants and affect groundwater quality.

QUESTION 198

Answer: C

Explanation: Ocean currents are primarily driven by the wind patterns on the Earth's surface, which influence the movement of water in the oceans.

QUESTION 199

Answer: C

Explanation: The epipelagic zone, also known as the sunlight zone, is the uppermost ocean layer where sunlight penetrates, supporting the highest biodiversity due to ample light availability for photosynthesis.

QUESTION 200

Answer: A

Explanation: Oxbow lakes are formed when a meandering river erodes its banks, creating a curved lake that was once a part of the river's course.

QUESTION 201

Answer: D

Explanation: A marsh is a type of freshwater wetland characterized by grasses, reeds, and shallow, standing water. Coral reefs, estuaries, and mangrove forests are coastal ecosystems, but not freshwater wetlands.

QUESTION 202

Answer: C

Explanation: Tides are primarily caused by the gravitational pull of the Moon and, to a lesser extent, the Sun on Earth's oceans.

QUESTION 203

Answer: C

Explanation: A mid-ocean ridge is formed when tectonic plates move apart, allowing magma to rise and create new oceanic crust.

QUESTION 204

Answer: A

Explanation: The salinity of seawater is influenced by the input of dissolved salts from river runoff, which carries minerals from the land into the ocean.

QUESTION 205

Answer: B

Explanation: Storm surges, which are caused by powerful storms like hurricanes, are a significant factor in coastal erosion due to their ability to push large amounts of water onto the shoreline.

QUESTION 206

Answer: D

Explanation: The Pacific Ocean is the largest and deepest ocean on Earth, covering a vast area and containing the Mariana Trench, the deepest point in any ocean.

QUESTION 207

Answer: D

Explanation: Coral bleaching occurs when corals expel their symbiotic algae due to stress, often caused by unusually high sea temperatures.

QUESTION 208

Answer: D

Explanation: Bogs are stagnant, acidic, and nutrient-poor freshwater ecosystems characterized by low oxygen levels and the accumulation of decaying organic matter.

QUESTION 209

Answer: C

Explanation: Barrier islands are formed by the accumulation of sand and sediment during storm surges and other wave events that deposit material along the coastline.

QUESTION 210

Answer: A

Explanation: The Gulf Stream is a warm ocean current that flows from the Gulf of Mexico along the eastern coast of North America before crossing the Atlantic Ocean towards Europe.

QUESTION 211

Answer: A

Explanation: Glaciers and ice caps store a significant portion of Earth's freshwater, and when they melt, they contribute to major water sources for human use.

QUESTION 212

Answer: B

Explanation: The primary factor influencing the salinity of a freshwater lake is the amount and type of precipitation it receives, which can vary depending on its location and climate conditions.

QUESTION 213

Answer: A

Explanation: The primary factor influencing the origin of ocean basins is the movement of tectonic plates on the Earth's surface. Plate tectonics play a crucial role in the formation of ocean basins through processes like seafloor spreading and subduction.

QUESTION 214

Answer: A

Explanation: Barrier islands are formed as a result of longshore drift, a process where sediments are moved along the coast by wave action. Over time, these sediments accumulate, forming barrier islands parallel to the mainland.

QUESTION 215

Answer: A

Explanation: The average salinity of ocean water worldwide is approximately 3.5%. This means that, on average, there are 35 grams of dissolved salts in every 1,000 grams of seawater.

QUESTION 216

Answer: A

Explanation: The Gulf Stream is responsible for carrying warm waters from the Gulf of Mexico along the eastern coast of the United States. This current has a significant impact on the region's climate, keeping it relatively mild and influencing weather patterns.

QUESTION 217

Answer: A

Explanation: Tides on Earth are primarily caused by the gravitational pull of the Moon and, to a lesser extent, the Sun. The gravitational attraction of these celestial bodies leads to the rise and fall of ocean water levels.

QUESTION 218

Answer: C

Explanation: Oil deposits are a non-living, abiotic marine resource. They are formed from the remains of ancient marine organisms and represent an essential geological resource.

QUESTION 219

Answer: D

Explanation: Fjord coastlines feature numerous winding inlets, often resembling the letter "W." These inlets are deep, glacially-carved valleys that are now filled with seawater.

QUESTION 220

Answer: C

Explanation: The Bathypelagic zone, also known as the midnight zone, receives the least amount of sunlight and is characterized by complete darkness. It is home to many bioluminescent organisms that produce light to navigate and communicate in the darkness.

QUESTION 221

Answer: D

Explanation: The Labrador Current brings cold waters from the Arctic southward along the eastern coast of Canada and the United States. It influences the climate of the region and plays a role in the formation of sea ice.

QUESTION 222

Answer: D

Explanation: Upwelling zones are biologically productive regions of the ocean where nutrient-rich waters from deeper layers are brought to the surface. This supports a high abundance of marine life and attracts various fish species.

QUESTION 223

Answer: A

Explanation: Mid-ocean ridges are underwater mountain chains formed by tectonic plate divergence. They are associated with volcanic activity and can be the site of hydrothermal vents.

QUESTION 224

Answer: A

Explanation: The primary cause of ocean currents is the wind. Wind-driven surface currents are a result of the frictional drag of moving air at the ocean's surface.

QUESTION 225

Answer: C

Explanation: Fish are considered a renewable and sustainable marine resource when managed properly through responsible fishing practices and regulations.

QUESTION 226

Answer: B

Explanation: Sea stacks are common erosional features found along rocky coastlines. They are formed when wave action erodes the softer rock around a more resistant rock, leaving behind a tall, isolated pillar.

QUESTION 227

Answer: B

Explanation: The Mesopelagic zone is known as the "twilight zone" and receives limited sunlight, making it difficult for photosynthesis to occur. This zone is characterized by a dim, blue light and houses various species of fish and other organisms adapted to low light conditions.

QUESTION 228

Answer: B

Explanation: The stratosphere contains the ozone layer, which plays a crucial role in absorbing harmful UV radiation, protecting life on Earth.

QUESTION 229

Answer: B

Explanation: Carbon dioxide is the most significant greenhouse gas in the atmosphere, contributing to the greenhouse effect and influencing global climate.

QUESTION 230

Answer: C

Explanation: Global wind patterns are primarily influenced by the Coriolis effect, a result of Earth's rotation, which deflects moving air masses.

QUESTION 231

Answer: A

Explanation: The troposphere is the layer closest to Earth's surface and is where most weather phenomena, such as clouds, storms, and precipitation, take place.

QUESTION 232

Answer: C

Explanation: The uneven heating of Earth's surface by the sun leads to temperature and pressure differences, which drive global wind patterns.

QUESTION 233

Answer: C

Explanation: Nitrogen dioxide is a key component of photochemical smog, formed by the reaction of nitrogen oxides with volatile organic compounds in the presence of sunlight.

QUESTION 234

Answer: B

Explanation: Chlorofluorocarbons (CFCs) are responsible for ozone layer depletion when they break down in the upper atmosphere, releasing chlorine atoms that destroy ozone molecules.

QUESTION 235

Answer: D

Explanation: Sublimation is the phase transition in which water vapor changes directly into ice without becoming a liquid first.

QUESTION 236

Answer: D

Explanation: The thermosphere contains the ionosphere, which is responsible for the auroras due to interactions with charged particles from the sun.

QUESTION 237

Answer: C

Explanation: El Niño refers to the warming of sea surface temperatures in the central and eastern Pacific Ocean, with widespread effects on weather patterns.

QUESTION 238

Answer: C

Explanation: Acid rain is primarily caused by the emission of sulfur dioxide and nitrogen oxides, which react with atmospheric moisture to form sulfuric and nitric acids.

QUESTION 239

Answer: C

Explanation: The greenhouse effect helps regulate Earth's temperature by trapping heat in the atmosphere, which is essential for maintaining a habitable climate.

QUESTION 240

Answer: A

Explanation: The troposphere is the layer where the temperature generally decreases with increasing altitude, except for the presence of a temperature inversion layer, which can trap pollutants.

QUESTION 241

Answer: C

Explanation: Solar radiation from the sun is the primary energy source that drives Earth's weather and climate systems.

QUESTION 242

Answer: C

Explanation: The rapid warming of the Earth's climate in recent times is primarily due to human activities, especially the burning of fossil fuels, leading to increased greenhouse gas emissions.

QUESTION 243

Answer: B

Explanation: The correct answer is the troposphere. It is the layer closest to the Earth's surface and contains most of the Earth's weather systems. Temperature generally decreases with altitude in the troposphere, making it conducive to vertical movement of air masses, leading to various weather phenomena.

QUESTION 244

Answer: B

Explanation: The correct answer is the Coriolis effect. It is the primary factor influencing global wind patterns by deflecting air masses towards the east or west depending on their hemisphere of origin. This effect, combined with differential heating of Earth's surface, leads to the formation of trade winds, westerlies, and polar easterlies.

QUESTION 245

Answer: D

Explanation: The correct answer is human activities, such as burning fossil fuels (coal, oil, and natural gas) and deforestation. These activities release significant amounts of greenhouse gases into the atmosphere, contributing to the enhanced greenhouse effect and global warming.

QUESTION 246

Answer: B

Explanation: The correct answer is that CFCs are commonly used in refrigeration and air conditioning systems as well as aerosol propellants. When released into the atmosphere, CFCs can reach the stratosphere, where they undergo photodissociation, releasing chlorine atoms that lead to ozone depletion.

QUESTION 247

Answer: B

Explanation: The correct answer is b. During an El Niño event, trade winds weaken, and the surface waters in the central and eastern Pacific Ocean become unusually warm. This warming disrupts normal weather patterns, leading to various impacts worldwide, including altered precipitation patterns, droughts, floods, and shifts in atmospheric circulation.

QUESTION 248

Answer: A

Explanation: High-pressure systems are associated with sinking air, which leads to stable atmospheric conditions. As the air sinks, it warms up, resulting in clear skies, warm temperatures, and light winds.

QUESTION 249

Answer: C

Explanation: Maritime tropical air masses form over warm tropical waters and are characterized by high temperatures and high humidity. When this air mass moves northward and affects the southeastern U.S., it brings hot and humid weather during the summer months.

QUESTION 250

Answer: D

Explanation: An occluded front occurs when a fast-moving warm front overtakes a slow-moving cold front, lifting the colder air off the ground and forming a complex weather system.

QUESTION 251

Answer: B

Explanation: Cumulonimbus clouds are tall and dense clouds that are associated with thunderstorms. They can bring heavy rain, lightning, and even severe weather such as tornadoes.

QUESTION 252

Answer: C

Explanation: The subtropical jet stream is located between the Ferrel and Polar cells in the upper troposphere. It is responsible for guiding weather systems and can impact the formation of weather patterns in mid-latitudes.

QUESTION 253

Answer: B

Explanation: The warm waters of the Gulf of Mexico can provide the necessary heat and moisture for the development and intensification of hurricanes in the Atlantic basin.

QUESTION 254

Answer: C

Explanation: A barometer is used to measure atmospheric pressure. It can help forecast changes in weather conditions by detecting shifts in air pressure.

QUESTION 255

Answer: B

Explanation: On a weather map, a blue line with half circles represents a cold front. This symbol indicates the boundary between advancing cold air and retreating warm air.

QUESTION 256

Answer: C

Explanation: Cumulus clouds are puffy and white with a flat base. They form when warm air rises and cools, causing water vapor to condense into visible cloud droplets.

QUESTION 257

Answer: B

Explanation: Sleet occurs when raindrops freeze into ice pellets before reaching the ground. It can create slippery conditions on roads and sidewalks.

QUESTION 258

Answer: A

Explanation: When air is forced to rise over a mountain range, it cools and loses moisture, resulting in a rain shadow effect on the leeward side. This creates a warmer and drier climate in that region.

QUESTION 259

Answer: B

Explanation: A tornado is a rapidly rotating column of air that extends from a thunderstorm to the ground. It is often associated with severe weather conditions and can cause significant damage.

QUESTION 260

Answer: A

Explanation: An anemometer is used to measure wind speed. It typically consists of cups that rotate with the wind, and the rate of rotation provides the wind speed measurement.

QUESTION 261

Answer: C

Explanation: Contour lines on a weather map connect points of equal air pressure. They help meteorologists analyze pressure patterns and identify high and low-pressure systems.

QUESTION 262

Answer: D

Explanation: Climate oscillations, such as El Niño and La Niña, refer to large-scale weather patterns that influence the climate of a region for an extended period. These patterns can have significant impacts on weather conditions, precipitation, and temperatures.

QUESTION 263

Answer: A

Explanation: Low-pressure systems are characterized by ascending warm, moist air, which cools and condenses as it rises. This process leads to the formation of towering cumulonimbus clouds that can produce thunderstorms. The unstable atmospheric conditions in low-pressure systems make them conducive to the development of severe weather, such as lightning, heavy rain, and occasionally tornadoes.

QUESTION 264

Answer: D

Explanation: Arctic air masses are characterized by their very cold temperatures and low moisture content. When these air masses move over land during winter, they bring bitterly cold and dry conditions to the central regions of a continent. The lack of moisture in these air masses leads to little or no precipitation during their passage, contributing to the dry winter weather in those areas.

QUESTION 265

Answer: C

Explanation: A stationary front occurs when a warm air mass and a cold air mass are moving toward each other but neither is displacing the other. As a result, the boundary between the two air masses remains nearly stationary or moves very slowly. Weather conditions along a stationary front can include extended periods of cloudiness and precipitation, as neither air mass is strong enough to displace the other and create significant weather changes.

QUESTION 266

Answer: B

Explanation: Cirrus clouds are high-altitude clouds that form at elevations above 20,000 feet. They are composed of ice crystals and are often associated with strong high-altitude winds. These clouds have a wispy appearance and tend to form in parallel bands or ribbons, indicating the direction and strength of the upper-level winds.

QUESTION 267

Answer: B

Explanation: Water has a higher heat capacity than land, meaning it can absorb and release heat more slowly. As a result, coastal regions near large bodies of water experience milder and more moderate climates compared to inland areas. During the day, the water's cooling effect keeps temperatures lower than inland regions, and at night, the water's warmth prevents temperatures from dropping as significantly as in inland areas. This moderation of temperature extremes creates a more comfortable climate in coastal regions.

QUESTION 268

Answer: C

Explanation: Carbon dioxide is the primary greenhouse gas that traps heat in the Earth's atmosphere, contributing to the greenhouse effect and influencing global climate.

QUESTION 269

Answer: B

Explanation: The ozone layer, which absorbs and filters harmful ultraviolet radiation from the Sun, is mainly located in the stratosphere.

QUESTION 270

Answer: B

Explanation: A barometer is used to measure atmospheric pressure, which helps in predicting weather changes and trends.

QUESTION 271

Answer: A

Explanation: The Coriolis effect is caused by the rotation of the Earth and affects the movement of air masses and ocean currents.

QUESTION 272

Answer: B

Explanation: Cumulonimbus clouds are large and vertically developed clouds that bring thunderstorms and heavy precipitation.

QUESTION 273

Answer: A

Explanation: A front is the boundary between two air masses with different properties, such as temperature and humidity, which can lead to weather changes.

QUESTION 274

Answer: C

Explanation: El Niño and La Niña are related to changes in ocean currents in the Pacific Ocean, affecting global weather patterns.

QUESTION 275

Answer: C

Explanation: Human activities, particularly the burning of fossil fuels, release large amounts of greenhouse gases, intensifying the natural greenhouse effect and contributing to global warming.

QUESTION 276

Answer: B

Explanation: The changing seasons on Earth are primarily caused by the tilt of the Earth's axis as it orbits the Sun.

QUESTION 277

Answer: C

Explanation: The troposphere is the lowest layer of the atmosphere, extending from the Earth's surface to an average height of about 10-12 kilometers.

QUESTION 278

Answer: A

Explanation: A tornado is a violent and dangerous weather phenomenon that forms from powerful thunderstorms and has a characteristic rotating column of air.

QUESTION 279

Answer: B

Explanation: Deserts are characterized by their hot and arid climate, low precipitation, and sparse vegetation.

QUESTION 280

Answer: A

Explanation: The two main components of the Earth's atmosphere are oxygen (about 21%) and nitrogen (about 78%).

QUESTION 281

Answer: C

Explanation: An anemometer is specifically designed to measure wind speed.

QUESTION 282

Answer: C

Explanation: Sublimation is the process by which water vapor transforms directly into ice without becoming liquid water first.

QUESTION 283

Answer: B

Explanation: Abiotic characteristics refer to non-living factors that influence the environment. Temperature fluctuations are a prime example of an abiotic factor as they can greatly impact the climate of a region.

QUESTION 284

Answer: B

Explanation: The El Niño/Southern Oscillation (ENSO) is a climate phenomenon characterized by the periodic warming and cooling of sea surface temperatures in the central and eastern equatorial Pacific Ocean, which can have significant global climate impacts.

QUESTION 285

Answer: C

Explanation: Climate change can lead to the disruption of migration patterns of various species, as changing environmental conditions can affect their habitats and food sources, impacting their traditional migration routes.

QUESTION 286

Answer: D

Explanation: Evaporation is a vital process in the hydrologic cycle, where water is converted from liquid to vapor form and rises into the atmosphere, eventually leading to condensation, precipitation, and the continuous flow of water through various Earth processes.

QUESTION 287

Answer: D

Explanation: Biotic characteristics refer to the living components of an ecosystem. Plant competition, where different plant species compete for resources, is an example of a biotic factor influencing a terrestrial ecosystem.

QUESTION 288

Answer: A

Explanation: The monsoon climate is primarily caused by Earth's axial tilt, which leads to differential heating of land and sea during different seasons, resulting in the seasonal reversal of wind patterns.

QUESTION 289

Answer: C

Explanation: Rising sea levels due to climate change pose significant threats to human society, as they can lead to coastal flooding, loss of land, and the displacement of populations living in vulnerable coastal areas.

QUESTION 290

Answer: C

Explanation: The tundra climate region experiences very low temperatures, minimal precipitation, and strong winds. The ground is often frozen, creating a challenging environment for vegetation to grow.

QUESTION 291

Answer: A

Explanation: In a positive feedback mechanism, a change in a system amplifies the initial change. Melting ice in polar regions can lead to a reduction in the Earth's albedo (reflectivity), causing more solar radiation to be absorbed, which, in turn, further accelerates ice melting and warming.

QUESTION 292

Answer: D

Explanation: Deserts are characterized by hot and dry conditions, and they typically support plants like cacti and other succulents that are adapted to conserve water in such arid environments.

QUESTION 293

Answer: D

Explanation: Ocean currents play a crucial role in redistributing heat around the Earth's surface. Warm ocean currents can warm coastal areas, while cold ocean currents can cool them, influencing regional climates.

QUESTION 294

Answer: D

Explanation: The thermohaline circulation, also known as the ocean conveyor belt, plays a role in regulating Europe's climate. Disruption of this circulation could lead to altered weather patterns in Europe, potentially affecting temperatures and precipitation.

QUESTION 295

Answer: C

Explanation: Fossil fuel combustion, such as burning coal, oil, and natural gas, is a major human activity that releases large amounts of greenhouse gases like carbon dioxide into the atmosphere, contributing to global warming and climate change.

QUESTION 296

Answer: C

Explanation: The greenhouse effect is the process in which certain gases in the Earth's atmosphere (like carbon dioxide and methane) trap heat from the sun, preventing it from escaping back into space. This phenomenon warms the Earth's surface, similar to how a greenhouse traps heat.

QUESTION 297

Answer: B

Explanation: The ozone layer in the stratosphere protects life on Earth from harmful ultraviolet (UV) radiation. A weakened ozone layer can result in more UV radiation reaching the Earth's surface, which could lead to an increased incidence of skin cancer and other health issues related to UV exposure.

QUESTION 298

Answer: A

Explanation: Nitrogen makes up about 78% of Earth's atmosphere, making it the most abundant gas.

QUESTION 299

Answer: B

Explanation: The stratosphere contains the ozone layer, which absorbs and blocks much of the Sun's harmful UV radiation.

QUESTION300

Answer: A

Explanation: The tropopause is the boundary that separates the troposphere and the stratosphere.

QUESTION 301

Answer: B

Explanation: Hurricanes are large rotating storm systems with low-pressure centers that form over warm ocean waters.

QUESTION 302

Answer: D

Explanation: Rain occurs when warm, moist air rises and cools, leading to condensation and precipitation.

QUESTION 303

Answer: B

Explanation: Cirrus clouds are thin, feathery clouds found at high altitudes and are composed of ice crystals.

QUESTION 304

Answer: A

Explanation: Carbon dioxide is a greenhouse gas that plays a significant role in trapping heat in the atmosphere, contributing to the greenhouse effect.

QUESTION 305

Answer: C

Explanation: Milankovitch cycles refer to changes in Earth's orbit and axial tilt, which influence the amount and distribution of solar radiation, impacting climate patterns.

QUESTION 306

Answer: C

Explanation: The ITCZ is an area where trade winds from both hemispheres converge, causing ascending air, cloud formation, and frequent precipitation.

QUESTION 307

Answer: A

Explanation: The troposphere is the layer where weather events primarily occur due to its proximity to the Earth's surface and the presence of moisture and temperature variations.

QUESTION 308

Answer: A

Explanation: Nitrogen and oxygen are the two most abundant gases in Earth's atmosphere, constituting almost all of its composition and supporting life as we know it.

QUESTION 309

Answer: A

Explanation: Fronts are boundaries between air masses with distinct characteristics. The clash of these air masses often leads to various weather phenomena like rain, thunderstorms, and shifts in wind direction.

QUESTION 310

Answer: A

Explanation: The greenhouse effect is a natural and vital process that helps maintain Earth's average temperature by trapping some of the outgoing heat, which keeps the planet warm enough to support life.

QUESTION 311

Answer: C

Explanation: Ocean currents are crucial in redistributing heat across the planet, affecting weather and climate patterns in coastal regions and beyond. Their role in heat transport has significant implications for regional and global climates.

QUESTION 312

Answer: A

Explanation: A solar eclipse occurs when the Moon comes between the Earth and the Sun, causing the Sun to be partially or completely obscured.

QUESTION 313

Answer: C

Explanation: Mars is often referred to as the "Red Planet" due to its reddish appearance, caused by iron oxide (rust) on its surface.

QUESTION 314

Answer: A

Explanation: The Kuiper Belt is a region in the outer solar system that contains numerous small icy bodies, including Pluto.

QUESTION 315

Answer: D

Explanation: The Sun is a star at the center of our solar system, providing light and heat to the planets orbiting around it.

QUESTION 316

Answer: D

Explanation: The changing phases of the Moon are a result of its position relative to the Sun and Earth, causing different amounts of the illuminated side to be visible from Earth.

QUESTION 317

Answer: B

Explanation: Saturn is famous for its extensive and beautiful ring system, composed of ice and rock particles.

QUESTION 318

Answer: B

Explanation: Titan, a moon of Saturn, is the largest moon in our solar system and is even larger than the planet Mercury.

QUESTION 319

Answer: B

Explanation: Johannes Kepler formulated three laws of planetary motion, which describe the elliptical orbits and the speed of planets around the Sun.

QUESTION 320

Answer: A

Explanation: The "Great Red Spot" is a massive storm on Jupiter, which has been observed for centuries.

QUESTION 321

Answer: C

Explanation: Nuclear fusion is the process by which stars, including our Sun, generate energy by fusing hydrogen nuclei to form helium.

QUESTION 322

Answer: A

Explanation: Aphelion refers to the point in a planet's orbit where it is farthest from the Sun.

QUESTION 323

Answer: B

Explanation: Our solar system is located in the Milky Way Galaxy, which is a spiral galaxy.

QUESTION 324

Answer: D

Explanation: When a massive star undergoes a supernova explosion and collapses, it can form a black hole with intense gravitational forces.

QUESTION 325

Answer: D

Explanation: Earth rotates around an imaginary line called its axis, causing the cycle of day and night.

QUESTION 326

Answer: C

Explanation: Europa, one of Jupiter's moons, is believed to have a subsurface ocean beneath its icy crust, making it an interesting target for the search for extraterrestrial life.

QUESTION 327

Answer: A

Explanation: The Sun's primary source of energy is nuclear fusion, where hydrogen nuclei combine to form helium in the core, releasing a tremendous amount of energy in the process.

QUESTION 328

Answer: D

Explanation: Main-sequence stars are in a stable phase of their life cycle, where gravitational forces pulling inwards are balanced by nuclear fusion reactions pushing outwards.

QUESTION 329

Answer: C

Explanation: During a supernova explosion, the intense energy and pressure cause the synthesis of elements heavier than iron through rapid neutron capture processes, also known as the r-process.

QUESTION 330

Answer: C

Explanation: Spiral galaxies are characterized by their flat, disk-like shape, with a central bulge and spiral arms extending outward from the center. Examples include the Milky Way Galaxy.

QUESTION 331

Answer: B

Explanation: The Big Bang Theory is the leading scientific explanation for the origin of the universe. It suggests that the universe began as a singularity and has been expanding and evolving ever since.

QUESTION 332

Answer: B

Explanation: Black holes have an incredibly strong gravitational pull that traps all matter and even light within their event horizon, making it impossible for anything to escape.

QUESTION 333

Answer: A

Explanation: Spectroscopy is a technique used to analyze the composition and properties of celestial objects by examining the different wavelengths of light they emit, which helps scientists understand their chemical composition and physical characteristics.

QUESTION 334

Answer: A

Explanation: Exoplanets are planets that orbit stars outside our solar system. The discovery and study of exoplanets have provided valuable insights into planetary systems beyond our own.

QUESTION 335

Answer: C

Explanation: Nuclear fusion is the process that powers the Sun and other stars. It involves the conversion of hydrogen into helium through the combination of atomic nuclei, releasing a tremendous amount of energy.

QUESTION 336

Answer: C

Explanation: Dark matter does not consist of ordinary matter like atoms and molecules. It is a hypothetical form of matter that does not emit light or interact with electromagnetic forces but interacts with gravity.

QUESTION 337

Answer: C

Explanation: A supernova is the final explosive stage in the life cycle of a massive star, during which it emits an enormous burst of light and energy before undergoing a collapse that can lead to the formation of a neutron star or a black hole.

QUESTION 338

Answer: A

Explanation: Quasars are distant and extremely luminous objects in space. Their brightness is attributed to the intense energy released as matter falls into a supermassive black hole at the core of a galaxy.

QUESTION 339

Answer: B

Explanation: The Hubble Space Telescope is renowned for its contributions to astronomy, providing stunning images and important data across multiple wavelengths, enhancing our understanding of the universe.

QUESTION 340

Answer: B

Explanation: When a massive star exhausts its nuclear fuel, it undergoes gravitational collapse, leading to the formation of a black hole, where the matter is concentrated in an infinitely small and dense point known as a singularity.

QUESTION 341

Answer: C

Explanation: Elliptical galaxies aremassive star exhausts its nuclear fuel, it undergoes gravitational collapse, leading to the formation of a black hole, where the matter is concentrated in an infinitely small and dense point known as a singularity.

QUESTION 342

Answer: C

Explanation: The periodic increase and decrease in brightness suggest that the star is being influenced by another celestial object, likely a companion star in a binary system. As the two stars orbit around their common center of mass, the varying position and orientation can cause changes in the observed brightness, leading to periodic fluctuations.

QUESTION 343

Answer: C

Explanation: A redshift in the spectral lines of a distant galaxy indicates that the light emitted by the galaxy is shifted towards longer wavelengths. This effect is due to the Doppler effect, and a redshift implies that the galaxy is moving away from Earth. The higher the redshift value, the faster the galaxy is receding.

QUESTION 344

Answer: B

Explanation: An object with an extremely high gravitational pull that light cannot escape from is indicative of a black hole. Black holes have such intense gravity that their event horizon prevents anything, including light, from escaping once it crosses the boundary.

QUESTION 345

Answer: D

Explanation: Luminosity is a measure of the total energy radiated by a star per unit time. It is independent of the observer's distance and gives an indication of how intrinsically bright a star is.

QUESTION 346

Answer: B

Explanation: A blue color in a galaxy's light indicates a higher proportion of young, hot stars emitting blue light. The presence of blue color suggests that the galaxy is currently experiencing a period of intense star formation, leading to the formation of young, massive, and hot stars.

QUESTION 347

Answer: B

Explanation: Mars is often referred to as the "Red Planet" because of its reddish color, which is caused by the presence of iron oxide (rust) on its surface.

QUESTION 348

Answer: C

Explanation: Jupiter is the largest planet in our solar system and is a gas giant composed mainly of hydrogen and helium.

QUESTION 349

Answer: B

Explanation: Comets are celestial objects composed of dust, rock, and ice. When they get close to the Sun, they develop a glowing tail due to the solar wind and radiation causing the ice to vaporize.

QUESTION 350

Answer: B

Explanation: Saturn is famous for its spectacular and extensive ring system, which is composed of ice particles and debris.

QUESTION 351

Answer: B

Explanation: Ganymede is the largest moon in our solar system and orbits the planet Jupiter. It is even larger than the planet Mercury.

QUESTION 352

Answer: D

Explanation: Saturn has the most rapid rotation among the listed options, completing one rotation in about 10 hours.

QUESTION 353

Answer: D

Explanation: The asteroid belt is a region between Mars and Jupiter where numerous irregularly shaped objects known as asteroids are found.

QUESTION 354

Answer: C

Explanation: Kepler's Third Law states that the square of a planet's orbital period is directly proportional to the cube of its average distance from the Sun.

QUESTION 355

Answer: C

Explanation: Olympus Mons is the largest volcano in our solar system and is located on the planet Mars.

QUESTION 356

Answer: A

Explanation: The "Great Red Spot" is a long-lived, high-pressure storm on Jupiter, which has been observed for centuries.

QUESTION 357

Answer: D

Explanation: Outgassing is the process by which a comet's icy nucleus vaporizes and forms a glowing atmosphere known as the coma when it gets close to the Sun.

QUESTION 358

Answer: A

Explanation: Venus has the most extreme temperatures in our solar system, with a thick atmosphere that traps heat, leading to scorching daytime temperatures and almost constant heat retention.

QUESTION 359

Answer: B

Explanation: The Cassini-Huygens spacecraft successfully landed on Saturn's moon Titan in 2005 and discovered lakes and seas of liquid hydrocarbons on its surface.

QUESTION 360

Answer: B

Explanation: Kepler's Second Law, also known as the Law of Equal Areas, states that a planet's orbital speed is faster when it is closer to the Sun and slower when it is farther away, resulting in equal areas being swept out in equal times.

QUESTION 361

Answer: A

Explanation: Ceres is the most massive asteroid in the asteroid belt between Mars and Jupiter and is alsoclassified as a dwarf planet.

QUESTION 362

Answer: C

Explanation: The formation of planetary rings is primarily due to the planet's strong gravitational forces. These forces capture and confine debris, such as rocks, dust, and ice particles, in orbit around the planet, forming the ring system. It is not solely based on the planet's proximity to the Sun.

QUESTION 363

Answer: B

Explanation: The "Moon Illusion" occurs because our brain interprets the moon's size relative to objects on the horizon, such as trees or buildings. When the moon is near the horizon, there are objects in our line of sight, which provides a visual reference, making the moon appear larger

QUESTION 364

Answer: A

Explanation: Venus has a thick atmosphere primarily composed of carbon dioxide, which creates a greenhouse effect. The thick atmosphere traps heat, preventing it from escaping back into space, leading to extreme surface temperatures on Venus, even though it is farther from the Sun than Mercury.

QUESTION 365

Answer: D

Explanation: Comets follow elliptical orbits due to the Sun's gravitational pull. When they come closer to the Sun (perihelion), the gravitational force is stronger, causing their orbits to elongate as they move away (aphelion). The gravitational interaction with other planets may influence the shape of a comet's orbit, but the primary reason for their elongation is the varying strength of the Sun's gravity along their path.

QUESTION 366

Answer: C

Explanation: The gas giants have massive sizes and strong gravitational forces. As a result, they were able to capture and retain a larger number of moons during their formation from the surrounding debris in the early solar system. The terrestrial planets are smaller and have weaker gravitational forces, leading to fewer captured moons.

QUESTION 367

Answer: A

Explanation: The leading scientific theory about the origin of the Moon is the Collision Theory, also known as the Giant Impact Hypothesis. It suggests that a celestial body, about the size of Mars, collided with Earth during the early formation of the solar system. The impact ejected a significant amount of debris into space, which eventually coalesced and formed the Moon.

QUESTION 368

Answer: D

Explanation: The primary reason for the Moon's tidal effects on Earth is its gravitational force. The Moon's gravitational pull creates two tidal bulges on Earth—one on the side facing the Moon and the other on the opposite side. These tidal bulges result in the regular rise and fall of ocean tides as Earth rotates within this gravitational field.

QUESTION 369

Answer: D

Explanation: The apparent path of the Sun across the sky over the course of a day is called the Ecliptic. It represents the Sun's apparent yearly motion against the backdrop of stars due to Earth's orbit around the Sun. The ecliptic is an important reference for understanding the motion of celestial bodies in the solar system.

QUESTION 370

Answer: C

Explanation: The tilt of Earth's axis is responsible for the changing seasons. As Earth orbits the Sun, different parts of the planet receive varying amounts of sunlight throughout the year. During summer in one hemisphere, that hemisphere is tilted towards the Sun, resulting in longer days and more direct sunlight. Conversely, during winter, the hemisphere is tilted away from the Sun, leading to shorter days and less direct sunlight.

QUESTION 371

Answer: A

Explanation: The phenomenon known as "precession" of Earth's axis is caused by the gravitational pull exerted by the Moon and the Sun on Earth's equatorial bulge. This gravitational force causes a slow, cyclic wobbling of Earth's rotational axis, completing approximately one full precession every 26,000 years.

QUESTION 372

Answer: A

Explanation: The apparent motion of planets relative to the stars from Earth's perspective is called "retrograde motion." Normally, planets move eastward (in the same direction as the stars) across the night sky. However, during certain periods, planets appear to reverse their motion temporarily and move westward, a phenomenon known as retrograde motion.

QUESTION 373

Answer: D

Explanation: The term for the point in the Moon's orbit where it is closest to Earth is "perigee." At perigee, the Moon is at its shortest distance from Earth, resulting in a slightly larger and brighter appearance in the sky, often referred to as a "supermoon."

QUESTION 374

Answer: B

Explanation: The stratosphere is the layer of the Earth's atmosphere that is responsible for absorbing most of the ultraviolet (UV) radiation from the Sun. Within the stratosphere, the ozone layer plays a crucial role in absorbing and filtering out harmful UV radiation, protecting life on Earth from its damaging effects.

QUESTION 375

Answer: A

Explanation: A solar eclipse occurs when the Moon passes directly between the Sun and Earth, causing the Moon to cast a shadow on Earth's surface. There are different types of solar eclipses, including total, partial, and annular, depending on the alignment and positioning of the Sun, Moon, and Earth.

QUESTION 376

Answer: C

Explanation: The phenomenon of "seasonal lag" is primarily caused by the heat absorption and release properties of the Earth's surface. After the summer solstice, the land and water bodies continue to absorb solar energy, leading to the warmest temperatures occurring a few weeks later. Similarly, after the winter solstice, the Earth's surface gradually releases stored heat, resulting in the coldest temperatures occurring a few

QUESTION 377

Answer: A

Explanation: Equatorial regions receive more direct sunlight throughout the year due to their proximity to the equator. The angle at which sunlight strikes the Earth's surface is nearly perpendicular, resulting in consistent and high solar energy input. As a result, these regions experience relatively stable temperatures without significant seasonal variations.

QUESTION 378

Answer: C

Explanation: The Moon appears to completely cover the Sun during a total solar eclipse because of a precise alignment of the Moon, Sun, and Earth. From the perspective of a viewer on Earth, the apparent sizes of the Moon and the Sun in the sky are similar due to their relative distances. When the Moon passes directly between the Sun and Earth in this configuration, it perfectly blocks the Sun's disk, resulting in a total solar eclipse.

QUESTION 379

Answer: A

Explanation: Tides and waves are both phenomena related to the movement of water bodies, but they have different causes and characteristics. Tides are primarily caused by the gravitational pull of celestial bodies, especially the Moon and to a lesser extent, the Sun. The regular rise and fall of tides occur due to the gravitational attraction of these celestial bodies on Earth's oceans, leading to the formation of tidal bulges. On the other hand, waves are generated by the transfer of energy from wind to the water surface, causing the water to move in a circular motion, but they do not have a regular pattern like tides.

QUESTION 380

Answer: A

Explanation: Retrograde motion, the apparent backward motion of planets relative to the stars, is observed from Earth due to the varying orbital speeds of Earth and the observed planet. Planets closer to the Sun, like Mercury and Venus, have minimal retrograde motion. However, planets with more noticeable retrograde motion, such as Mars, Jupiter, and Saturn, are farther from the Sun. As the relative positions of these planets and Earth change during their orbits, their apparent motion against the backdrop of stars appears to temporarily reverse, creating the retrograde motion effect.

QUESTION 381

Answer: B

Explanation: The greenhouse effect is a natural process that helps regulate the Earth's temperature. Certain greenhouse gases, such as carbon dioxide (CO_2) and methane (CH_4), present in the Earth's atmosphere, trap and absorb outgoing infrared radiation (heat) from the Earth's surface. This prevents the heat from escaping directly into space and leads to the warming of the lower atmosphere and Earth's surface. Without the greenhouse effect, Earth would be much colder, making it difficult to support life as we know it. However, human activities have been contributing to an increase in greenhouse gas concentrations, leading to enhanced greenhouse effect and global warming.

QUESTION 382

Answer: C

Explanation: The color of a star is directly related to its surface temperature. Hotter stars appear blue or white, while cooler stars appear red.

QUESTION 383

Answer: D

Explanation: Galaxies come in various shapes, such as spiral, elliptical, and irregular. Scientists classify galaxies based on these shapes.

QUESTION 384

Answer: C

Explanation: The Sun is the central and most massive component of the solar system, providing light, heat, and gravitational influence to all other objects in the system.

QUESTION 385

Answer: C

Explanation: The Sun is a ball of hot, ionized gas (plasma) and does not have a solid surface. It consists of different layers, such as the core, radiative zone, and convective zone.

QUESTION 386

Answer: B

Explanation: A solar eclipse occurs when the Moon's shadow falls on the Earth's surface, blocking out the Sun either partially or completely.

QUESTION 387

Answer: B

Explanation: Stars and planets appear to move across the sky due to the rotation of the Earth on its axis and the orbital motion of the Earth around the Sun. The stars also have their own motions within the galaxy.

QUESTION 388

Answer: D

Explanation: The ecliptic is the apparent path that the Sun takes across the celestial sphere as seen from Earth over the course of a year.

QUESTION 389

Answer: D

Explanation: Opposition is the position of a planet when it is on the opposite side of the Earth from the Sun. During opposition, the planet rises as the Sun sets, and it is visible throughout the night.

QUESTION 390

Answer: C

Explanation: A star's mass is the most significant factor determining its life cycle, including its duration, stages (like main sequence, red giant, etc.), and eventual fate (white dwarf, neutron star, black hole, etc.).

QUESTION 391

Answer: B

Explanation: The Sun generates energy through nuclear fusion, where hydrogen atoms fuse together to form helium, releasing an enormous amount of energy in the process.

QUESTION 392

Answer: C

Explanation: Neutron stars are the remnants of massive stars that have undergone a supernova explosion. They are incredibly dense, with a mass greater than that of the Sun but compressed into a sphere about the size of a city.

QUESTION 393

Answer: A

Explanation: A nova is a sudden increase in the brightness of a star due to a thermonuclear explosion on its surface. It does not lead to the destruction of the star.

QUESTION 394

Answer: C

Explanation: Comets are celestial objects composed mainly of dust, ice, and gas. When they approach the Sun, the solar heat causes the ices to vaporize and form a bright coma and tail.

QUESTION 395

Answer: B

Explanation: The Milky Way Galaxy is the spiral galaxy that contains our solar system along with billions of other stars, planets, and celestial objects.

QUESTION 396

Answer: B

Explanation: Cosmology is the scientific study of the origin, evolution, and overall structure of the universe, including its galaxies, stars, and planets.

QUESTION 397

Answer: A

Explanation: In general, red stars tend to be cooler than blue or white stars. The color of a star is directly related to its surface temperature. Cooler stars emit more red light, while hotter stars emit more blue and white light.

QUESTION 398

Answer: A

Explanation: The Moon does not emit its own light; it reflects sunlight that falls on its surface. During a lunar eclipse, the Earth blocks direct sunlight from reaching the Moon, but some sunlight is scattered and refracted by the Earth's atmosphere, allowing a portion of it to reach the Moon, making it visible.

QUESTION 399

Answer: B

Explanation: According to Kepler's laws of planetary motion, planets farther from the Sun have longer orbital periods. Therefore, a planet located beyond Neptune would take more time to complete one orbit around the Sun compared to Earth's orbital period.

QUESTION 400

Answer: A

Explanation: Absolute magnitude is a measure of a star's intrinsic brightness, assuming it is located at a standard distance from Earth. If two stars have the same absolute magnitude, but one appears dimmer from Earth, it means the dimmer star is farther away, as its light has to travel a longer distance to reach us.

QUESTION 401

Answer: A

Explanation: Density is the mass of an object divided by its volume. Since both planets have the same mass and size, the planet made of iron, being a denser material than water ice, would have a higher density. Iron has a higher mass per unit volume than water ice, leading to a higher overall density.

QUESTION 402

Answer: D

Explanation: Reviewing current knowledge in light of existing evidence is a crucial principle for designing and conducting an inquiry-based scientific investigation. It involves understanding and evaluating the existing body of knowledge related to the research question before proposing new answers, explanations, or predictions.

QUESTION 403

Answer: C

Explanation: As an Earth science teacher, you can model scientific attitudes like curiosity and openness to new ideas by encouraging students to explore and ask questions. Creating an environment that fosters curiosity and openness is essential for promoting active learning and inquiry-based methods in science instruction.

QUESTION 404

Answer: C

Explanation: Designing the learning environment to support student inquiries and scientific investigations involves providing ample time, space, and resources for all students to participate. This approach ensures that every student has an opportunity to engage in fieldwork, laboratory experiments, and other forms of scientific investigation.

QUESTION 405

Answer: C

Explanation: A primary role of the Earth science teacher is to guide students in generating, refining, focusing, and testing scientific questions and hypotheses. Teachers play a crucial role in facilitating the inquiry process and helping students develop their scientific thinking and problem-solving skills.

QUESTION 406

Answer: C

Explanation: Active learning and inquiry-based methods are essential in science instruction because they promote deeper understanding and critical thinking among students. By actively engaging in scientific inquiries, students develop problem-solving skills, explore concepts in-depth, and develop a more profound appreciation for the scientific process.

QUESTION 407

Answer: C

Explanation: Option C best exemplifies guiding students to identify, refine, and focus scientific ideas and questions. By facilitating a class discussion, the teacher encourages students to actively engage with their data and observations, encouraging them to refine their initial explanations. Through this discussion, students can compare their findings, consider alternative explanations, and identify potential sources of error, thus promoting higher-level thinking skills and scientific problem-solving.

QUESTION 408

Answer: C

Explanation: Option C aligns best with the teacher's objectives. The classroom debate will help uncover common misconceptions and encourage students to critically analyze the theory of plate tectonics. The individual research projects will allow students to delve deeper into specific tectonic events, expanding their understanding of the topic. Finally, the student-led presentations will challenge them to apply their knowledge, use higher-level thinking skills, and present their findings to their peers.

QUESTION 409

Answer: D

Explanation: Option D would further support the students' inquiry-based investigation. By encouraging students to record unexpected or unusual observations, the teacher promotes the development of scientific curiosity and critical thinking. These observations may lead to new questions, hypotheses, or insights, allowing students to refine and focus their scientific ideas. Additionally, discussing these observations as a group will help students to analyze and evaluate different explanations for their results, fostering scientific problem-solving skills.

QUESTION 410

Answer: C

Explanation: Option C would best help the teacher move students from concrete to more abstract understanding. A class discussion allows students to actively engage with the topic, share their ideas, and build upon their prior knowledge. It encourages higher-level thinking skills and logical reasoning as they explore the impact of human activities on natural resources and the concept of sustainability. By discussing real-world applications and the importance of sustainable practices, students are more likely to develop a deeper and more abstract understanding of renewable and non-renewable resources.

QUESTION 411

Answer: B

Explanation: The correct strategy is to encourage students to brainstorm independently, allowing them to generate their own scientific questions related to plate tectonics. This approach promotes critical thinking, creativity, and individual exploration of the topic. It helps students refine and focus their ideas and questions, contributing to a more comprehensive inquiry-based scientific investigation.

QUESTION 412

Answer: C

Explanation: To help students analyze different explanations for the volcanic eruption, the best approach is to present multiple explanations and promote a classroom discussion where students can compare and evaluate them. This strategy allows students to engage in higher-level thinking, logical reasoning, and scientific problem-solving as they critically assess the strengths and weaknesses of each explanation.

QUESTION 413

Answer: D

Explanation: To guide students in making systematic observations and measurements during their inquiry-based investigation, Ms. Lee should provide clear instructions and tools that are relevant to studying soil erosion. This ensures that students have a structured approach to data collection, allowing them to gather consistent and meaningful information that can be analyzed effectively.

QUESTION 414

Answer: C

Explanation: To uncover common misconceptions among students and address them effectively, Mr. Anderson should encourage an open and inclusive classroom environment where students feel comfortable sharing their ideas about climate change. This approach allows the teacher to identify misconceptions and facilitates discussions that challenge students to expand their understanding of the topic. It also helps build upon their prior knowledge and correct any inaccurate information.

QUESTION 415

Answer: C

Explanation: Rubrics are an excellent tool for evaluating students' participation in and understanding of inquiry-based scientific investigations. They provide clear criteria for assessment and can be used to assess various aspects of the project, such as the quality of work, problem-solving skills, data analysis, and communication. Unlike multiple-choice or true/false questions, which are more limited in scope, rubrics offer a comprehensive assessment of students' performance in a project-based setting.

QUESTION 416

Answer: B

Explanation: Mr. Smith should design a performance assessment to match the learning objectives of the unit on plate tectonics. Performance assessments allow students to demonstrate their knowledge and skills in a hands-on manner, such as creating models, conducting experiments, or analyzing real-world data related to plate tectonics. This type of assessment aligns well with inquiry-based learning and encourages deeper understanding and critical thinking, unlike short-answer or true/false questions, which may not capture the students' full comprehension of the subject.

QUESTION 417

Answer: D

Explanation: To continually monitor students' understanding of science concepts and skills on an ongoing basis, Ms. Lee should use formative assessment methods. Formative assessments provide feedback and help identify students' strengths and areas for improvement during the learning process. Unlike summative assessments that occur at the end of a unit or semester, formative assessments are more frequent and enable teachers to adjust their instructional practices accordingly. Self-assessment is beneficial for students to reflect on their learning but may not be the most effective method for teachers to continuously monitor their students' understanding.

QUESTION 418

Answer: C

Explanation: Limiting the use of an assessment to its intended purpose is crucial to maintaining fairness and validity. If an assessment is used for purposes other than what it was designed for, it may not accurately measure what it is supposed to assess. This can lead to unfair evaluations of students' performance and misinterpretation of their understanding. Additionally, using assessments for unintended purposes may compromise the validity and reliability of the assessment results. It is essential for teachers to use assessments that align with their specific objectives to ensure accurate and meaningful evaluation of students' learning.

QUESTION 419

Answer: A

Explanation: Earth Science teachers should plan activities that are inclusive and consider the interests, knowledge, understanding, abilities, and experiences of all students, including English-language learners. Providing inquiry-based investigations relevant to students' daily lives helps foster engagement and motivation in their own learning. This strategy allows students to connect scientific concepts to their real-world experiences and makes the learning process more meaningful and enjoyable.

QUESTION 420

Answer: A

Explanation: Earth Science teachers should consider research-based factors such as students' prior knowledge, experience, and attitudes when understanding how they develop scientific understanding. These factors significantly influence science learning, as students' existing knowledge and attitudes can impact their engagement and receptiveness to new concepts and ideas.

QUESTION 421

Answer: A

Explanation: Earth Science teachers can promote student self-motivation and engagement by designing instructional materials that use situations from students' daily lives. By connecting scientific content to their real-world experiences, students are more likely to be motivated to learn and actively engage with the subject matter.

QUESTION 422

Answer: A

Explanation: Earth Science teachers should use a variety of instructional strategies to ensure all students comprehend content-related texts. Different students have varying learning preferences and abilities, and employing diverse instructional approaches can cater to these individual differences. Utilizing a range of texts and technologies will help students locate, retrieve, and retain information effectively.

QUESTION 423

Answer: B

Explanation: Conducting an interactive group discussion allows the Earth Science teacher to gauge students' prior knowledge and misconceptions effectively. During the discussion, the teacher can ask probing questions, observe students' responses, and identify common misconceptions. This approach encourages student participation and provides valuable insights into their existing understanding of science concepts.

QUESTION 424

Answer: D

Explanation: Evaluating the consistency of scores obtained from multiple administrations of the assessment helps determine its reliability. If the assessment consistently produces similar results over different occasions, it is more likely to be reliable. Additionally, comparing the scores of the assessment with external measures or established criteria helps determine its validity.

QUESTION 425

Answer: C

Explanation: Sharing evaluation criteria with students before the assessment allows them to understand the expectations and standards they are being assessed against. This empowers students to engage in meaningful self-assessment, as they can reflect on their work in the context of the criteria provided, identify areas of improvement, and take ownership of their learning.

QUESTION 426

Answer: B

Explanation: A take-home assignment with flexible deadlines allows all students, regardless of their learning pace or personal circumstances, to have equal opportunities to demonstrate their achievements. Unlike a multiple-choice test with a time limit, which may cause stress and hinder some students' performance, a take-home assignment provides a more accommodating environment for students to showcase their understanding of Earth Science concepts.